Astrophysics and Space Science Proceedings

Volume 38

For further volumes:
http://www.springer.com/series/7395

Claudia Moreno González
José Edgar Madriz Aguilar
Luz Marina Reyes Barrera
Editors

Accelerated Cosmic Expansion

Proceedings of the Fourth International Meeting On Gravitation and Cosmology

Springer

Editors
Claudia Moreno González
Universidad de Guadalajara
Guadalajara, Jalisco
México

José Edgar Madriz Aguilar
Universidad de Guadalajara
Guadalajara, Jalisco
México

Luz Marina Reyes Barrera
Universidad de Guadalajara
Guadalajara, Jalisco
México

ISSN 1570-6591
ISBN 978-3-319-02062-4 ISBN 978-3-319-02063-1 (eBook)
DOI 10.1007/978-3-319-02063-1
Springer Cham Heidelberg New York Dordrecht London

Library of Congress Control Number: 2013955583

Printed on acid-free paper

Springer is part of Springer Science+Business Media (www.springer.com)

Preface

This volume is a collection of papers representing the proceedings of the IV International Meeting on Gravitation and Cosmology, which took place on May 21-25, 2012, in the city of Guadalajara, Jalisco, México.

This meeting was sponsored by ICTP-Trieste (Italy) and COECyTJAL—Universidad de Guadalajara (México). The conference is part of a series of scientific meetings devoted to current and selected topics in gravitation and cosmology. These meetings began at the Universidad Central Marta Abreu de la Villas in Santa Clara, Cuba, in 2004.

The goal of this meeting was to attract the attention of leading experts in the fields of gravitation and cosmology. The subjects discussed provided both an update and an evaluation of the state of alternative theories of gravity, in connection with the issue of the accelerating expansion of the universe. Topics reviewed included f(R) theories, dark matter and dark energy issues, Modified Newtonian Dynamics (MOND) models, scalar tensor theories derived from non-Riemannian geometries, emergent universes and the cosmological constant. Attendees included younger and senior researchers as well as graduate students. All contributions in this volume have been refereed by the scientific committee.

Finally, we would like to thank to all those who helped us to make this meeting such a success. In particular, we would like to express our grateful thanks to Professor Israel Quiros for bringing the event to the Universidad de Guadalajara. Special thanks goes out as well to the authorities of the Universidad de Guadalajara for their support during the meeting.

Guadalajara, Jalisco, México
August 2013

Claudia Moreno González
José Edgar Madriz Aguilar
Luz Marina Reyes Barrera

Acknowledgements

The Abdus Salam International Centre for Theoretical Physics ICTP, Trieste
Dr. George Thompson
Consejo Estatal De Ciencia Y Tecnología De Jalisco COECyTJAL
Dr. Francisco Medina Gómez
Universidad De Guadalajara Centro Universitario De Ciencias Exactas E Ingenierías CUCEI
Dr. Cesar Octavio Monzón
Museo Regional De Guadalajara Gobierno Del Estado De Jalisco, México
Mtro. Francisco Javier De La Peña Anguiano

Committee

Organizing Committee
Dra. Claudia Moreno González[1]
Dr. José Edgar Madriz Aguilar[2]
Dra. Luz Marina Reyes Barrera
Dr. Miguel Angel Olmos Gómez

Scientific Committee
Dr. Mauricio Bellini
Dra. Nora Bretón
Dr. Jorge Cervantes
Dr. Francisco Guzmán
Dr. Xavier Hernández
Dra. Ruth Lazkoz
Dr. Roy Maartens
Dr. Shin'ichi Nojiri
Dr. Darío Núñez
Dr. Octavio Obregón
Dr. Sergei Odintsov
Dr. Diego Pavon
Dr. Carlos Romero
Dr. Emmanuel Saridakis
Dr. Roberto Sussman
Dr. Gerardo Torres del Castillo
Dr. Shani Varun
Dr. David Wands

Local Organizing Committee
Dr. Juan Carlos Degollado Daza
Dr. Israel Quiros Rodríguez
Dr. Ricardo García Salcedo
M. en C. Pablo Torres Tonche

---

[1] claudia.moreno@cucei.udg.mx
[2] madrizaguilar@yahoo.com.mx

Technical Support
Rodolfo Ibarra Nuño
Luis Miguel Borruel Chavira
Laura Lorena Razón Gallegos
Quetzalcóatl Ornelas García
M. en C. Victor Guadalajara Estrada

Contents

Contributors

C. Allen Instituto de Astronomía, Universidad Nacional Autónoma de México, Distrito Federal, México

T. S. Almeida Departamento de Física, Universidade Federal da Paraíba, Jõao Pessoa, Brasil

Diego A. Ballén-Daza Escuela de Física, Universidad Industrial de Santander, Bucaramanga, Colombia

Nora Bretón Dpto. de Física, Centro de Investigación y de Estudios Avanzados del I. P. N., D.F., México

Salvatore Capozziello Dipartimento di Scienze Fisiche, Universitá di Napoli "Federico II" and INFN Sez. di Napoli, Compl. Univ. di Monte S. Angelo, Napoli, Italy

Carlos Custodio Dpto. de Física, Centro de Investigación y de Estudios Avanzados del I. P. N., D.F., México

Josue De-Santiago Institute of Cosmology & Gravitation, University of Portsmouth, Portsmouth, Reino Unido

Valerio Faraoni Physics Department and STAR Research Cluster, Bishop's University, Sherbrooke, Canada

Guillermo A. González Escuela de Física, Universidad Industrial de Santander, Bucaramanga, Colombia

E. I. Guendelman Physics Department, Ben Gurion University of the Negev, Beer Sheva, Israel

Antonio C. Gutiérrez-Piñeres Instituto de Ciencias Nucleares, UNAM, Mexico, Colombia

X. Hernandez Instituto de Astronomía, Universidad Nacional Autónoma de México, Distrito Federal, México

A. Jiménez Instituto de Astronomía, Universidad Nacional Autónoma de México, Distrito Federal, México

Pedro Labraña Departamento de Física, Universidad del Bío-Bío, Casilla 5–C, Concepción, Chile

Tonatiuh Matos Departamento de Física, Centro de Investigación y de Estudios Avanzados del I. P. N., D.F., México

Diego Pavón Departamento de Física, Universidad Autónoma de Barcelona, Barcelona, Spain

M. L. Pucheu Departamento de Física, Universidade Federal da Paraíba, Jõao Pessoa, Brasil

Ninfa Radicella Dipartimento di Física, Università di Salermo, Salermo, Italy

Tanja Rindler-Daller Department of Astronomy and Texas Cosmology Center, The University of Texas at Austin, Austin TX, USA

Victor H. Robles Departamento de Física, Centro de Investigación y de Estudios Avanzados del I. P. N., D.F., México

C. Romero Departamento de Física, Universidade Federal da Paraíba, Jõao Pessoa, Brasil

Paul R. Shapiro Department of Astronomy and Texas Cosmology Center, The University of Texas at Austin, Austin TX, USA

Cosimo Stornaiolo Istituto Nazionale di Fisica Nucleare, Complesso Universitario di Monte S. Angelo, Napoli, Italy

Abril Suárez Departamento de Física, Centro de Investigación y de Estudios Avanzados del I. P. N., D.F., México

Roberto A Sussman Instituto de Ciencias Nucleares, UNAM, México, México

David Wands Institute of Cosmology & Gravitation, University of Portsmouth, Portsmouth, Reino Unido

Yuting Wang Institute of Cosmology & Gravitation, University of Portsmouth, Portsmouth, Reino Unido

Part I
Modified Theories of Gravity

Chapter 1
Recovering Flat Rotation Curves and Galactic Dynamics From f(*R*)-Gravity

Salvatore Capozziello

Abstract The so-called $f(R)$-gravity addresses issues like dark energy and dark matter from the point of view of gravitational field instead of requiring new material ingredients. In particular, several self-gravitating structures, like spiral and elliptical galaxies, can be consistently described, in the weak-field limit regime, due to the presence of Yukawa-like corrections in the Newtonian potential. We give here some examples related to the observations.

1.1 Introduction

Dark matter issues come from dynamical mass estimates of self-gravitating systems. In several astrophysical observations, there is more matter dynamically inferred than that can be accounted for from luminous components. This mass discrepancy is usually attributed to additional (missing or dark) matter, assuming the validity of Newton law of gravity at astrophysical scales. Oort was the first that posed the 'missing matter' problem [1]. By observing the Doppler red-shift values of stars moving near the plane of our Galaxy, he asserted that he could calculate how fast the stars were moving. He found that there had to be enough matter inside the galaxy such that the central gravitational force was strong enough to keep the stars from escaping, much as Sun's gravitational pull keeps a planet in its orbit. But when the calculation was made, it turned out that there was not enough mass in the Galaxy. The discrepancy was not small: the Galaxy had to be at least twice as massive as the sum of the mass of all its visible components combined. In addition, in the 1960's the radial profile of the tangential velocity of stars in their orbits around the Galactic Center, as a function of their distance from that center, was measured. It was found that typically, once we get away from the Galactic Center, all the stars travel with the same velocity independent of their distance out from the Galactic Center.

There were problems, too, at a larger scale. In 1933, Zwicky announced that when he measured the individual velocities of a large group of galaxies, known as the Coma cluster, he found that all of the galaxies that he considered were moving

S. Capozziello (✉)
Dipartimento di Scienze Fisiche, Universitá di Napoli "Federico II" and INFN Sez. di Napoli, Compl. Univ. di Monte S. Angelo, Edificio G, Via Cinthia, I-80126, Napoli, Italy
e-mail: capozzie@na.infn.it

C. Moreno González et al. (eds.), *Accelerated Cosmic Expansion*, Astrophysics and Space Science Proceedings 38,
DOI 10.1007/978-3-319-02063-1_1, © Springer International Publishing Switzerland 2014

so rapidly relative to one another that the cluster should have come apart long ago. The visible mass of the galaxies making up the cluster was far too little to produce enough gravitational force to hold the cluster together. So not only our own galaxy was lacking mass, but also the whole Coma cluster of galaxies was suffering the same problem at a different scale [2]. Initially, the problem was only approached by leaving Newton's law inviolated and postulating the existence of some invisible dark entities to make up the missing mass. At the beginning, it has never came to mind anyone to go back and examine the basic assumption that only gravity was at work in these cases. It was easier to patch up the theory introducing invisible entities. Many names have been coined to define this invisible entity, a bit as in the days of ether.

There are the MAssive Compact Halo Objects (MACHOs), objects like black holes, and neutron stars that purportedly populate the outer reaches of galaxies like the Milky Way. Then there are the Weakly Interacting Massive Particles (WIMPs), which possess mass, yet do not interact with ordinary matter (baryons such as protons and neutrons) because they are composed by something unknown. Dark (missing) matter (DM) even comes in two flavors, hot (HDM) and cold (CDM). The CDM is supposedly to be in dead stars, planets, brown dwarfs ("failed stars") etc., while HDM is postulated to be fast moving in particles floating throughout the universe. It should be constituted by neutrinos, tachyons etc. But where is all of this missing matter? The truth is that after many years of looking for it, there is still no definitive proof that WIMPs exist, or that MACHOs will ever make up more than five percent of the total reserve of missing dark stuff. Besides, by adding a further ingredient, the cosmological constant Λ, such a model (now ΛCDM) has become the new cosmological paradigm usually called the *concordance model*. In fact, high quality data coming from the measurements of cluster properties as the mass, the correlation function and the evolution with redshift of their abundance, the Hubble diagram of Type Ia Supernovae, the optical surveys of large scale structure, the anisotropies of the cosmic microwave background, the cosmic shear measured from weak lensing surveys and the Lyman-α forest absorption are evidences toward a spatially flat universe with a subcritical matter content and undergoing a phase of accelerated expansion. Interpreting all this information in a self-consistent model is the main task of modern cosmology and ΛCDM model provides a good fit to the most part of the data giving a reliable picture of the today observed universe.

Nevertheless, it is affected by serious theoretical shortcomings that have motivated the search for alternative candidates generically referred to as *dark energy* or *quintessence*. Such models range from scalar fields rolling down self interaction potentials to phantom fields, from phenomenological unified models of dark energy and dark matter to alternative gravity theories.

Essentially, dark energy (or any alternative component) has to act as a negative pressure fluid which gives rise to an overall acceleration of the Hubble fluid. Despite of the clear mechanisms generating the observed cosmological dynamics, the nature and the fundamental properties of dark energy remain essentially unknown notwithstanding the great theoretical efforts made up to now.

The situation for dark matter is similar: its clustering and distribution properties are fairly well known at every scale but its nature is unknown, up to now, at fundamental level.

On the other hand, the need of unknown components as dark energy and dark matter could be considered nothing else but as a signal of the breakdown of Einstein General Relativity (GR) at astrophysical (galactic and extragalactic) and cosmological scales. In this context, Extended Theories of Gravity (ETGs) could be, in principle, an interesting alternative to explain cosmic acceleration and large scale structure without any missing components. In their simplest version, the Ricci curvature scalar R, linear in the Hilbert-Einstein action, could be replaced by a generic function $f(R)$ whose true form could be "reconstructed" by the data. In fact, there is no a priori reason to consider the gravitational Lagrangian linear in the Ricci scalar while observations and experiments could contribute to define and constrain the "true" theory of gravity [3]. Coming to the weak-field limit, which essentially means considering Solar System scales, any alternative relativistic theory of gravity is expected to reproduce GR which, in any case, is firmly tested only in this limit and at these scales [4]. Even this limit is a matter of debate since several relativistic theories do not reproduce exactly the Einstein results in their Newtonian limit but, in some sense, generalize them.

In general, all these efforts can be included in the so called Modified Newtonian Dynamics (MOND), first proposed by Milgrom [5], in the attempt to explain missing matter without dark matter but assuming a change into dynamics at scales larger than Solar System's ones. In general, any relativistic theory of gravitation yields corrections to the weak-field gravitational potentials which, at the post-Newtonian (PN) level and in the Parametrized Post-Newtonian (PPN) formalism, could constitute a test of these theories [4]. This point deserves a deep discussion. Beside the fundamental physics motivations coming from Quantum Gravity and unification theories [3], ETGs pose the problem that there are further gravitational degrees of freedom (related to higher order terms, non-minimal couplings and scalar fields in the field equations) and gravitational interaction is *not* invariant at any scale. This means that, besides the Schwarzschild radius, other characteristic gravitational scales come out from dynamics. Such scales, in the weak field approximation, should be responsible of characteristic lengths of astrophysical structures that should result *confined* in this way.

Here, without claim of completeness, we will try to address the problem to describe some self-gravitating structures without dark matter asking for corrections to the Newtonian potential that could fit data and reproduce dynamics. In this sense, the MOND approach can be reproduced not in a phenomenological way but starting from a field theory [6]. It is possible to show that *any analytic ETG, except GR, presents Yukawa-like corrections in the weak-field limit*. From an astrophysical point of view, these corrections means that further scales have to be taken into account and that their effects could be irrelevant at local scales as Solar System. The emergence of Yukawa-like corrections to the Newtonian potential is discussed in Sect. 1.2 where the weak-field limit of $f(R)$-gravity, the simplest ETG, is worked out. Here, $f(R)$ is a generic function of the Ricci curvature scalar R. A large class of $f(R)$

models are compatible with results of Solar System experiments and constraints coming from Equivalence Principle, as shown in [7]. These results are extremely important because the philosophy is that ETGs are not *alternative* to GR but are just an extension of it that could take into account the whole dynamics of gravitational field. In fact, as soon as $f(R) = R$, GR is fully recovered. In Sect. 1.3, we use the modified gravitational potential to reproduce the rotation curves of spiral galaxies by using simulated data and confronting with the Navarro-Frenk-White model for the halo. Furthermore, we extend the test to elliptical galaxies considering long-slit and planetary nebulae data. It is found that the Yukawa-like potential is able to fit nicely these elliptical galaxies and the anisotropy distribution is consistent with that estimated if a dark halo is considered. The parameter which measures the "strength" of the Yukawa-like correction is, on average, smaller than the one found for spiral galaxies and correlates both with the scale length of the Yukawa-like term and the orbital anisotropy. Conclusions are drawn in Sect. 1.4.

1.2 Yukawa-Like Corrections from the Weak-Field Limit of f(R)-Gravity

The simplest ETG in four dimensions is given by the action

$$S = \int d^4x\sqrt{-g}\left[f(R) + \mathcal{L}^{(m)}\right], \tag{1.1}$$

where $f(R)$ is an unspecified function of scalar curvature R. The term $\mathcal{L}^{(m)}$ is the minimally coupled ordinary matter contribution, considered as a *perfect fluid*. This is the straightforward extension of GR as soon as it is assumed $f(R) \neq R$. Varying with respect to $g_{\mu\nu}$, we get

$$\begin{aligned} G_{\alpha\beta} &= \frac{1}{f'(R)}\left\{\frac{1}{2}g_{\alpha\beta}\left[f(R) - Rf'(R)\right] + f'(R)_{;\alpha\beta} - g_{\alpha\beta}\Box f'(R)\right\} + \frac{T^{(m)}_{\alpha\beta}}{f'(R)} \\ &= T^{(curv)}_{\alpha\beta} + \frac{T^{(m)}_{\alpha\beta}}{f'(R)}, \end{aligned} \tag{1.2}$$

where $T^{(curv)}_{\alpha\beta}$ is an effective stress-energy tensor constructed by the extra curvature terms. We are considering physical units for the gravitational coupling. In the case of GR, $T^{(curv)}_{\alpha\beta}$ identically vanishes and standard, minimal coupling is recovered for matter contribution. In order to deal with standard self-gravitating systems, any theory of gravity has to be developed to its Newtonian or post-Newtonian limit depending on the order of approximation of the theory in terms of power of c^{-2}, where c is the speed of light [8].

In principle, the following analysis can be developed for any ETGs but we will consider only the $f(R)$ case. As discussed in [3], we can deal with the Newtonian and the post-Newtonian limit of $f(R)$ gravity adopting the spherical symmetry.

The solution of field equations can be obtained considering the general spherically symmetric metric:

$$ds^2 = g_{\sigma\tau}dx^\sigma dx^\tau = g_{00}(x^0, r)dx^{02} - g_{rr}(x^0, r)dr^2 - r^2 d\Omega \tag{1.3}$$

where $x^0 = ct$ and $d\Omega$ is the solid angle. In order to develop the Newtonian limit, let us consider the perturbed metric with respect to a Minkowskian background $g_{\mu\nu} = \eta_{\mu\nu} + h_{\mu\nu}$. The metric entries can be developed as:

$$\begin{cases} g_{tt}(t,r) \simeq 1 + g_{tt}^{(2)}(t,r) + g_{tt}^{(4)}(t,r), \\ g_{rr}(t,r) \simeq -1 + g_{rr}^{(2)}(t,r), \\ g_{\theta\theta}(t,r) = -r^2, \\ g_{\phi\phi}(t,r) = -r^2 \sin^2\theta, \end{cases} \tag{1.4}$$

where we are assuming $c = 1$, $x^0 = ct \to t$ and expansion is at order c^{-2} and c^{-4}. Since we want to obtain the most general result, we does not provide any specific form for $f(R)$-Lagrangian. We assume, however, analytic Taylor expandable $f(R)$ functions with respect to a certain value $R = R_0$:

$$f(R) = \sum_n \frac{f^n(R_0)}{n!}(R - R_0)^n \simeq f_0 + f_1 R + f_2 R^2 + f_3 R^3 + \cdots . \tag{1.5}$$

In order to obtain the weak field approximation, one has to insert expansions (1.4) and (1.5) into field Eqs. (1.2) and expand the system up to the orders $\mathcal{O}(0)$, $\mathcal{O}(2)$ e $\mathcal{O}(4)$ (that is, as stated above at order c^{-2} and c^{-4}). This approach provides general results and specific (analytic) theories are selected by the coefficients f_i in Eq. (1.5). It is worth noticing that, at the order $\mathcal{O}(0)$, the field equations give the condition $f_0 = 0$ and then the solutions at further orders do not depend on this parameter. If we consider the $\mathcal{O}(2)$-order approximation, the field equations in vacuum, results to be

$$\begin{cases} f_1 r R^{(2)} - 2 f_1 g_{tt,r}^{(2)} + 8 f_2 R_{,r}^{(2)} - f_1 r g_{tt,rr}^{(2)} + 4 f_2 r R^{(2)} = 0, \\ f_1 r R^{(2)} - 2 f_1 g_{rr,r}^{(2)} + 8 f_2 R_{,r}^{(2)} - f_1 r g_{tt,rr}^{(2)} = 0, \\ 2 f_1 g_{rr}^{(2)} - r\left[f_1 r R^{(2)} - f_1 g_{tt,r}^{(2)} - f_1 g_{rr,r}^{(2)} + 4 f_2 R_{,r}^{(2)} + 4 f_2 r R_{,rr}^{(2)}\right] = 0, \\ f_1 r R^{(2)} + 6 f_2 \left[2 R_{,r}^{(2)} + r R_{,rr}^{(2)}\right] = 0, \\ 2 g_{rr}^{(2)} + r\left[2 g_{tt,r}^{(2)} - r R^{(2)} + 2 g_{rr,r}^{(2)} + r g_{tt,rr}^{(2)}\right] = 0. \end{cases} \tag{1.6}$$

It is evident that the trace equation (the fourth in the system (1.6)), provides a differential equation with respect to the Ricci scalar which allows to solve exactly the system (1.6) at $\mathcal{O}(2)$-order. Finally, one gets the general solution:

$$\begin{cases} g_{tt}^{(2)} = \delta_0 - \frac{Y}{f_1 r} - \frac{\delta_1(t)e^{-r\sqrt{-\xi}}}{3\xi r} + \frac{\delta_2(t)e^{r\sqrt{-\xi}}}{6(-\xi)^{3/2}r} \\ g_{rr}^{(2)} = -\frac{Y}{f_1 r} + \frac{\delta_1(t)[r\sqrt{-\xi}+1]e^{-r\sqrt{-\xi}}}{3\xi r} - \frac{\delta_2(t)[\xi r+\sqrt{-\xi}]e^{r\sqrt{-\xi}}}{6\xi^2 r} \\ R^{(2)} = \frac{\delta_1(t)e^{-r\sqrt{-\xi}}}{r} - \frac{\delta_2(t)\sqrt{-\xi}e^{r\sqrt{-\xi}}}{2\xi r} \end{cases}$$

where $\xi \doteq \frac{f_1}{6f_2}$, and Y is an arbitrary integration constant. When we consider the limit $f(R) \to R$, in the case of a point-like source of mass M, we recover the standard Schwarzschild solution. Let us notice that the integration constant δ_0 is dimensionless, while the two arbitrary functions of time $\delta_1(t)$ and $\delta_2(t)$ have respectively the dimensions of $length^{-1}$ and $length^{-2}$. The functions of time $\delta_i(t)$ $(i = 1, 2)$ are completely arbitrary since the differential equation system (1.6) contains only spatial derivatives and can be fixed to constant values. Besides, the integration constant δ_0 can be set to zero since it represents an unessential additive quantity for the potential. It is possible to write the general solution of the problem considering the previous expression (1.3). In order to match at infinity the Minkowskian prescription for the metric, one can discard the Yukawa growing mode and then we have:

$$\begin{cases} ds^2 = \left[1 - \frac{2GM}{f_1 r} - \frac{\delta_1(t)e^{-r\sqrt{-\xi}}}{3\xi r}\right] dt^2 \\ -\left[1 + \frac{2GM}{f_1 r} - \frac{\delta_1(t)(r\sqrt{-\xi}+1)e^{-r\sqrt{-\xi}}}{3\xi r}\right] dr^2 - r^2 d\Omega \\ R = \frac{\delta_1(t)e^{-r\sqrt{-\xi}}}{r} \end{cases} \tag{1.7}$$

At this point, one can provide the gravitational potential. The first of (1.7) gives the second order solution in term of the metric expansion (see the definition (1.4)). This term coincides with the gravitational potential at the Newton order. In particular, since $g_{tt} = 1 + 2\Phi_{grav} = 1 + g_{tt}^{(2)}$, the gravitational potential of $f(R)$-gravity, analytic in the Ricci scalar R, is

$$\Phi_{grav} = -\left[\frac{GM}{f_1 r} + \frac{\delta_1(t)e^{-r\sqrt{-\xi}}}{6\xi r}\right]. \tag{1.8}$$

This general result means that the standard Newton potential is achieved only in the particular case $f(R) = R$ while it is not so for any analytic $f(R)$ models. Equation (1.8) deserves some comments. The parameters $f_{1,2}$ and the function δ_1 represent the deviations with respect the standard Newton potential. To test these theories of gravity inside the Solar System, we need to compare such quantities with respect to the current experiments, or, in other words, Solar System constraints should be evaded fixing such parameters. On the other hand, these parameters could acquire non-trivial values (e.g. $f_1 \neq 1$, $\delta_1(t) \neq 0$, $\xi \neq 1$) at scales different from the Solar System ones. We note that the ξ parameter can be related to an effective mass being

$m^2 = (3\xi)^{-1}$ and can be interpreted also as an effective length L. Equation (1.8) can be recast as

$$\Phi(r) = -\frac{GM}{(1+\delta)r}\left(1+\delta e^{-\frac{r}{L}}\right), \tag{1.9}$$

where the first term is the Newtonian-like part of the potential for a point-like mass $\frac{M}{1+\delta}$ and the second term is a modification of the gravity including a scale length L associated to the above coefficients of the Taylor expansion. If $\delta = 0$ the Newtonian potential and the standard gravitational coupling are recovered. Comparing Eqs. (1.8) and (1.9), we obtain that $1+\delta = f_1$, and δ is related to $\delta_1(t)$ through

$$\delta_1 = -\frac{6GM}{L^2}\left(\frac{\delta}{1+\delta}\right) \tag{1.10}$$

where $\frac{6GM}{L^2}$ and δ_1 can be assumed quasi-constant. Under these assumptions, the scale length L could naturally reproduce several phenomena that range from Solar System to cosmological scales. In the following section, we show examples of self-gravitating systems where this potential can be used to address the missing matter issues. Understanding on which scales the modifications to GR are working or what is the weight of corrections to the Newtonian gravitational potential is a crucial point that could confirm or rule out these alternative approaches.

1.3 The Cases of Spiral and Elliptical Galaxies

As we said, ETGs can impact on the estimate of missing matter properties at galactic scales. In fact, corrections on the gravitational potential give rise to modifications on the rotation curve [9]. The amount of such modifications can be compared with Newtonian mechanics and then constrains the halo model parameters by fitting the theoretical rotation curve to the observed one. In other words, one can see how modified gravity could be a possible way to solve the cusp/core and similar problems related to the presence of dark matter.

Here we take into account spiral and elliptical galaxies but also larger scales, as galaxy clusters, can be considered within this approach [10]. Let us assume a gravitational potential as Eq. (1.9). It is worth noticing that a Yukawa - like correction has been used several times in the past showing that the observed flat rotation curves of spiral galaxies may be well fitted by this model with no need for dark haloes provided $\delta < 0$ and L adjusted on a case-by-case basis [11]. Equation (1.9) is the starting point for the computation of the rotation curve of an extended system.

One has to remember that, in the Newtonian gravity, the circular velocity in the equatorial plane is given by $v_c^2(R) = Rd\Phi/dR|_{z=0}$, with Φ the total gravitational potential. Thanks to the superposition principle and the linearity of the point mass potential on the mass, this latter is computed by adding the contribution from infinitesimally small mass elements and then transforming the sum into an integral over the mass distribution. For a spherically symmetric source, one can simplify this procedure considering the Gauss Theorem to find out that the usual result $v_c(r) = GM(r)/r$ with $M(r)$ the total mass within r. However, because of the Yukawa - like correction, the Gauss Theorem does not apply anymore and hence one must generalize the derivation of the gravitational potential. Alternatively, one can remember that $v_c(R) = RF(R, z = 0)$ being F the total gravitational force and R a radial coordinate [12]. This is the starting point where a general expression can be derived for the case of a generic potential giving rise to a separable force, *i.e.*:

$$F_p(\mu, r) = \frac{GM_\odot}{r_S^2} f_\mu(\mu) f_r(\eta), \tag{1.11}$$

with $\mu = M/M_\odot$, $\eta = r/r_S$ and $(M_\odot, r_s)$ the Solar mass and a characteristic length of the problem. In our case, it is $f_\mu = 1$ and:

$$f_r(\eta) = \left(1 + \frac{\eta}{\eta_L}\right) \frac{\exp\left(-\eta/\eta_L\right)}{(1+\delta)\eta^2}, \tag{1.12}$$

with $\eta_L = L/r_S$. Using cylindrical coordinates (R, θ, z) and the corresponding dimensionless variables (η, θ, ζ) (with $\zeta = z/r_S$), the total force then reads:

$$F(\mathbf{r}) = \frac{G\rho_0 r_S}{1+\delta} \int_0^\infty \eta' d\eta' \int_{-\infty}^\infty d\zeta' \int_0^\pi f_r(\Delta) \tilde{\rho}(\eta', \theta', \zeta') d\theta', \tag{1.13}$$

with $\tilde{\rho} = \rho/\rho_0$, ρ_0 a reference density, and we can define

$$\Delta = \left[\eta^2 + \eta'^2 - 2\eta\eta' \cos(\theta - \theta') + (\zeta - \zeta)'^2\right]^{1/2}. \tag{1.14}$$

For obtaining axisymmetric systems, one can set $\tilde{\rho} = \tilde{\rho}(\eta, \zeta)$. Furthermore, the systems we are considering here are spiral galaxies which will be modeled as the sum of an infinitesimally thin disc and a spherical halo, and then the scaling radius r_s will be the disc scale length R_d. Under these assumptions, the rotation curve may then be obtained as:

$$v_c^2(R) = \frac{G\rho_0 R_d^2 \eta}{1+\delta} \int_0^\infty \eta' d\eta' \int_{-\infty}^\infty \tilde{\rho}(\eta', \zeta') d\zeta' \int_0^\pi f_r(\Delta_0) d\theta', \tag{1.15}$$

with

$$\Delta_0 = \Delta(\theta = \zeta = 0) = \left[\eta^2 + \eta'^2 - 2\eta\eta' \cos\theta' + \zeta'^2\right]^{1/2}. \tag{1.16}$$

Inserting Eq. (1.12) into Eq. (1.15) gives rise to an integral which has to be evaluated numerically even for the spherically symmetric case. It is evident that the total rotation curve may be split in the sum of the standard Newtonian term and a corrective one disappearing for $L \to \infty$, *i.e.* when ETGs have no deviations from GR at galactic scales. Assuming that a spiral galaxy can be modeled as the sum of an infinitesimally thin disc and a spherical halo and denoting with **p** the halo model parameters, it is possible to calculate the total rotation curve as:

$$v_c^{2(R,M_d,\mathbf{p}_i)} = v_{dN}^{2(R,M_d)} + v_{hN}^2(R,\mathbf{p}_i) + v_{dY}^{2(R,M_d)} + v_{hY}^2(R,\mathbf{p}_i),$$

where M_d is the disc mass, the labels d and h denote disc and halo related quantities, while N and Y refer to the Newtonian and Yukawa - like contributions [9]. This means that one may model a spiral galaxy as the sum of a thick disc and a spherical halo without dark matter contribution. Therefore to simulate the rotation curves, one can refer

- to spiral galaxies with reasonable values of the model parameters and
- the sampling and the noise should be the same as actual data.

In Fig. 1.1 it is shown an example of simulated curves with superimposed theoretical curves.

It is generated the disc scalelength R_d and the halo virial mass from flat distributions over the ranges:

$$0.5 \leq R_d/R_{d,MW} \leq 2.0 \quad 11.5 \leq \log M_{vir} \leq 13.5$$

with $R_{d,MW} = 2.55$ kpc, the disc scalelength for the Milky Way [13]. In order to set the disc mass, we first define the halo mass fraction (DM) within the optical radius as:

$$f_{DM} = \frac{M_h(R_{opt})}{M_d + M_h(R_{opt})}, \tag{1.17}$$

with $R_{opt} = 3.2R_d$ the optical radius. The disc mass within R_d can be approximated with the total disc mass. One can generate f_{DM} from a flat distribution in the range $(0.9, 1.1)f_{DM,fid}$ and $f_{DM,fid} = 50\%$ (see, *e.g.*, [14] and refs. therein). The halo scale-length R_s is computed as $R_s = R_{vir}/c$ where the concentration c is randomly generated from a Gaussian centered on:

$$c = 16.7\left(\frac{M_{vir}}{10^{11}\ h^{-1}\ \mathrm{M}_\odot}\right)^{-0.125}, \tag{1.18}$$

and variance set to 10 % of the mean value. Note that the above relation has been derived by [15] for the mass range $(0.03, 30) \times 10^{12}\mathrm{M}_\odot$ following the method detailed in [16] and updating the cosmological model. Finally, it is necessary to set the modified potential parameters (δ, L) and so one first can set $\delta = 1/3$ and run different simulations randomly generating $\log \eta_L = \log(\lambda/R_d)$ from a flat distribution covering the wide range $(-2, 2)$. In order to explore the impact of δ, one can also consider the extremal case $\delta = 1.0$ thus maximizing the contribution of the Yukawa term. The sample of simulated rotation curves is the starting point of the analysis. Indeed, one can fit each curve with a given (Newtonian) model and compare the output best fit parameters with the input ones [9]. In realistic situations, one has a set of (R, v_c) data

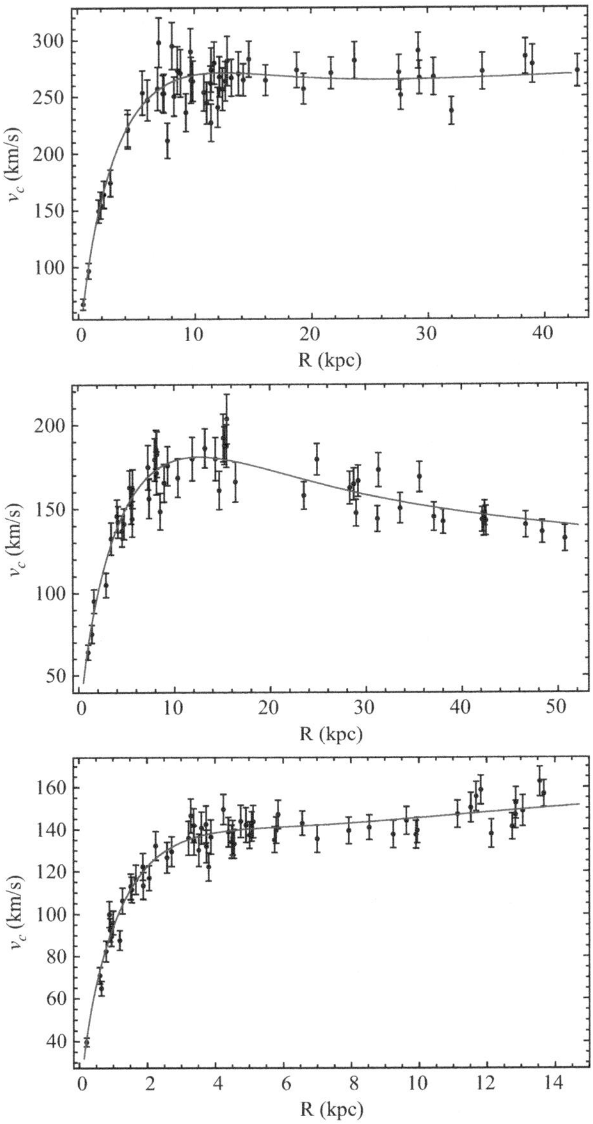

Fig. 1.1 Examples of simulated rotation curves with superimposed theoretical curves. From left to right, model parameters are $(\log M_d, \log M_{vir}, c, f_{DM}, \log \eta_L) = (11.15, 12.90, 10.24, 0.47, 0.36)$, $(10.90, 11.76, 14.77, 0.45, -0.92)$, $(10.04, 12.10, 13.76, 0.54, 1.11)$, while the simulation parameters are set as discussed in the text. Note that, depending on how the model parameters are set, it is possible to get rotation curves which are flat, decreasing or increasing in the outer region [9]

and a measurement of the galaxy surface brightness in a given band. It is then common to set the disc scale-length R_d to the value inferred from photometry, while the disc mass can be inferred from the total luminosity, provided an estimate of the stellar $M/\mathcal{L}$ ratio is somewhat available (e.g. from stellar population synthesis models fitted to the galaxy colors). As a first step, it is assumed that both (R_d, M_d) are known and fit the simulated data for each rotation curve to determine the Navarro-Frenk-White

Table 1.1 Bias $\mathcal{R}(x)$ on the NFW model parameters assuming the disc mass is known and $\delta = 1/3$. Columns are as follows: 1. parameter id; 2., 3., 4. mean $\pm$ standard deviation, median and rms values, 5., 6., 7., 8., 9. Spearman rank correlation coefficients between $\mathcal{R}(x)$ and input $\log \eta_s$, $\log M_{vir}$, c, f_{DM}, $\log \eta_\lambda$. Upper half of the table is for the well fitted sample, while lower half for the best fit sample one

Id	$\langle \mathcal{R} \rangle$	$\mathcal{R}_{med}$	$\mathcal{R}_{rms}$	$C(\log \eta_s, \mathcal{R})$	$C(\log M_{vir}, \mathcal{R})$	$C(c, \mathcal{R})$	$C(f_{DM}, \mathcal{R})$	$C(\log \eta_\lambda, \mathcal{R})$
η_s	4.0 ± 2.8	2.9	4.8	0.31	0.18	−0.17	−0.09	−0.47
M_{vir}	3.9 ± 3.8	2.4	5.5	0.48	0.31	−0.27	−0.05	−0.32
c	0.45 ± 0.15	0.45	0.48	−0.13	−0.05	0.07	0.13	0.55
f_{DM}	0.70 ± 0.06	0.70	0.71	0.36	0.30	−0.20	0.39	0.70
η_s	3.8 ± 2.8	2.8	4.7	0.32	0.24	−0.23	−0.07	−0.42
M_{vir}	3.6 ± 3.6	2.2	5.1	0.50	0.36	−0.34	−0.02	−0.26
c	0.47 ± 0.15	0.47	0.49	−0.14	−0.10	0.13	0.12	0.53
f_{DM}	0.70 ± 0.06	0.70	0.71	0.39	0.31	−0.18	0.40	0.71

(NFW) model parameters only. Table 1.1 gives some statistics on the distribution of $\mathcal{R}(x)$ for the halo parameters (η_s, M_{vir}). However, these values could be misleading since the histograms for $\mathcal{R}(\eta_s)$ and $\mathcal{R}(M_{vir})$ are strongly asymmetric with long tails towards the right. In other words, $\mathcal{R}(\eta_s)$ and $\mathcal{R}(M_{vir})$ could be much larger than their median value with $\mathcal{R}(\eta_s)$ being as high as 12 and values of $\mathcal{R}(M_{vir})$ as large as 15 so that both the halo scale-length and virial mass can be grossly overestimated (see [9] for details). Since $c \propto R_{vir} \propto M_{vir}^{1/3}$, one could naively expect that the concentration is overestimated too. On the contrary, the distribution of the $\mathcal{R}(c)$ values is almost symmetric around its mean clearly disfavoring $\mathcal{R}(c) > 1$, *i.e.* the concentration is underestimated. Actually, such a result can be understood remembering that $c \propto R_s^{-1}$ so that $\mathcal{R}(c) \propto \mathcal{R}^{1/3}(M_{vir})/\mathcal{R}(\eta_s)$ finally leading to values smaller than unity. The emerging picture is therefore that of a halo having a larger mass than the input one, but also a larger scale-length. This can be qualitatively explained noting that the Yukawa terms in the rotation curve increase the net circular velocity both in the inner and outer regions probed by the data. In order to adjust the fit in the halo dominated regions, one has to increase M_{vir}, but then R_s has to be increased too in order to lower the density (and hence the contribution to the rotation curve) in the inner regions not to overcome the observed circular velocity. Albeit a detailed analysis is required, one can finally stress that this analysis points towards a new usage of the rotation curves to be considered with and without assuming the presence of dark matter.

The modified potential (1.9) can be tested also for elliptical galaxies checking whether it is able to provide a reasonable match to their kinematics. Such self-gravitating systems are very different with respect to spirals so addressing both classes of objects under the same standard could be a fundamental step versus dark matter [17]. One may construct equilibrium models based on the solution of the radial Jeans equation to interpret the kinematics of planetary nebulae [17]. One can use the inner long slit data and the extended planetary nebulae kinematics for three galaxies which have published dynamical analyses within well-defined dark matter halo framework. They are: NGC 3379, NGC 4494, NGC 4374 (see [17] and references therein). The decreasing velocity dispersion profiles of the first two galaxies have been modeled with an intermediate mass halo, $\log M_{\rm vir} \sim 12 - 12.2 M_\odot$, with concentration $c_{\rm vir} =$

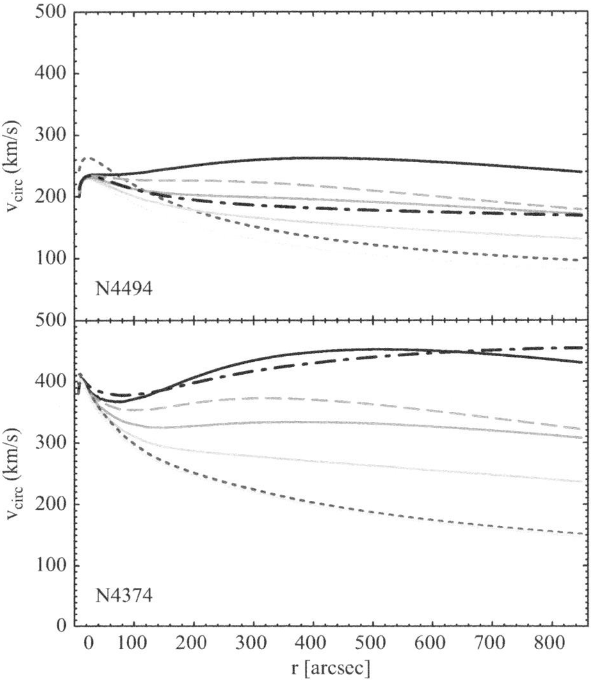

Fig. 1.2 Circular velocity produced by the modified potential for the two galaxies N4494 (*top*) and N4374 (*bottom*). In both cases the $M/\mathcal{L}_*$ has been fixed to some fiducial value (as expected from stellar population models and Kroupa 2001 IMF): $M/\mathcal{L}_* = 4.3\Upsilon_{\odot,B}$ for NGC 4494 and $M/\mathcal{L}_* = 5.5\Upsilon_{\odot,V}$ for NGC 4374. The potential parameters adopted are: $L = 250''$ and $\delta = 0, -0.65, -0.8, -0.9$ (lighter to darker solid lines) and $L = 180''$ and $\delta = -0.8$ (dashed lines). The dotted line is a case with positive coefficient of the Yukawa-like term and $L = 5000''$ which illustrates that positive δ cannot produce flat circular velocity curves. Finally some reference Navarro-Frenk-White models [20] are shown as dot-dashed lines [17]

$6 - 8$ and a fair amount of radial anisotropy in the outer regions. For NGC 4374, having a rather flat dispersion profile, a more massive (adiabatically contracted) halo with $\log M_{\rm vir} \sim 13.4 M_{\odot}$ and $c_{\rm vir} \sim 7$ is required with a negligible amount of anisotropy in the outer regions. These models turn out to be in fair agreement with the expectation of WMAP5, $c_{\rm vir} - M_{\rm vir}$ relation and with the Initial Mass Function discussed in [18]. In [17], it is shown that this sample is particularly suitable for a comparison with alternative theories of gravity. In Fig. 1.2, it is shown the circular velocity of the modified potential as a function of the potential parameters L and δ for NGC 4494 and NGC 4374. As for the spiral galaxies, negative values of the δ parameter make the circular velocity more and more flat also reproducing the typical dip (e.g. NGC 4374) of the circular velocity found for the dark matter models (dot-dashed curves) of the most massive systems. On the contrary, positive δ values do not produce easily flat circular velocity curves. From a theoretical viewpoint, this is not a problem since, as discussed above, δ is a free parameter that can assume positive and negative values. However, comparing results for spirals and ellipticals, it is clear that the morphology of these two classes of systems strictly depends on the sign and the value of δ.

A consistency check with galaxy scaling relations as Tully-Fisher relation for spirals and Faber-Jackson relation for ellipticals can also be considered [17]. From the model point of view, the problem of fitting a modified potential as in Eq. (1.9) implies the same kind of degeneracies between the anisotropy parameter, $\beta = 1 - \sigma_\theta^2/\sigma_r^2$ (where σ_θ and σ_r are the azimuthal and radial dispersion components in spherical coordinates), and the non-Newtonian part of the potential (characterized by two parameters like typical dark haloes) in a similar way of the classical mass-anisotropy degeneracy. It can be shown that these degeneracies can be alleviated via higher-order Jeans equations including, in the dynamical models, both the dispersion ($\sigma_{\rm p}$) and the kurtosis (κ) profiles of the tracers. Under spherical assumption, no-rotation and $\beta =$ const (corresponding to the family of distribution functions $f(E,\mathcal{L}) = f_0\mathcal{L}^{-2\beta}$, see [19] and references therein, the 2nd and 4th moment radial equations can be compactly written as:

$$s(r) = r^{-2\beta}\int_r^\infty x^{2\beta} H(x)dx\,, \tag{1.19}$$

where $s(r) = \{\rho\sigma_r^2; \rho\overline{v_r^4}\}$, β is the anisotropy parameter, and

$$H(r) = \left\{\rho\frac{d\Phi}{dr}; 3\rho\frac{d\Phi}{dr}\overline{v_r^2}\right\}, \tag{1.20}$$

respectively for the dispersion and kurtosis equations, being the latter $\kappa(r) = \overline{v_r^4}/\sigma_r^4$. Equation (1.19) are the ones interested by the potential modification and include four free parameters to be best-fitted: the $f(R)$ parameters $\{\delta, L\}$, the "dynamically inferred" stellar mass-to-light ratio. In Fig. 1.3, we show the dispersion and kurtosis profiles of the three galaxies with the $f(R)$ models superimposed (solid lines). The fitting procedure is based on the simultaneous χ^2 minimization of the dispersion and kurtosis profiles over a regular grid in the parameter space. The best-fit parameters are summarized in Table 1.2 together with some information on the galaxy sample. The overall agreement of model curves with data is remarkably good and it is comparable with models obtained with dark matter modeling (gray lines in Fig. 1.3). In all cases, the $f(R)$-gravity models allow to accommodate a constant orbital anisotropy β which is very close to the estimates derived from the corresponding dark matter models (see Table. 1.2). This is mainly guaranteed by the fit to the $\kappa(r)$ which does not respond much differently to the modified potential with respect the dark matter models. Thus, an important result of the analysis is that the orbital anisotropy is fairly stable to the change of the galaxy potential.

A final comment is in order at this point. These considerations lead to the conclusion that potential (1.9) can work, in principle, for very different self-gravitating systems. However, the value and the sign of its parameters can be different, depending on the secondary features of the systems. Comparing the further length coming from the Yukawa correction to the other fundamental gravitational length, the Schwarzschild radius, it seems that the additional degrees of freedom, coming from ETGs, could rule also the structures and morphologies of systems not only the whole amount of mass.

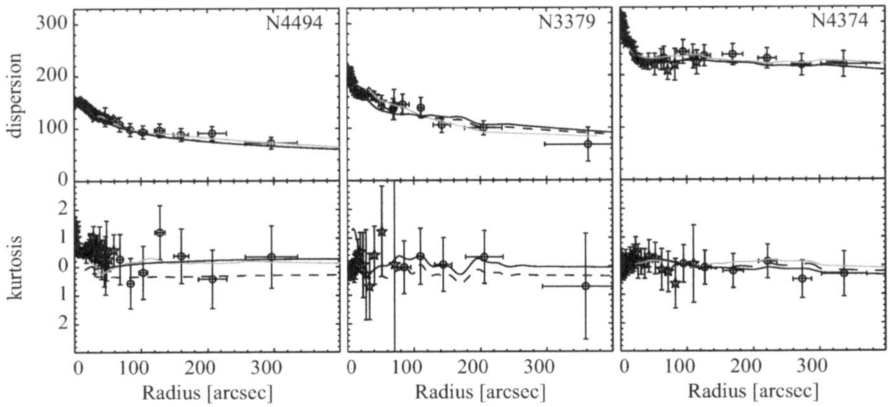

Fig. 1.3 Dispersion in *kms* (top) and kurtosis (bottom) fit of the galaxy sample for the different $f(R)$ parameter sets: the anisotropic solution (solid lines) is compared with the isotropic case (dashed line – for NGC 4374 and NGC 4494 this is almost indistinguishable from the anisotropic case). From the left, NGC 4494, NGC 3379 and NGC 4374 are shown with missing matter models as gray lines from N+09, DL+09 (no kurtosis is provided), and N+11 respectively [17].

Table 1.2 Model parameters for the Yukawa potential. Galaxy ID, total magnitude, effective radius and model parameters for the unified solution. DM-based estimates for $M/\mathcal{L}_\star$ and β (NGC 3379: DL+09; NGC 4374: N+11; NGC 4494: N+09) are shown in parentheses for comparison. $M/\mathcal{L}_\star$ are in solar units, R_{eff} and L in kpc. Typical errors on $M/\mathcal{L}_\star$ are of the order of $0.2M/\mathcal{L}_\odot$ and on β of 0.2). The small χ^2 values are mainly due to the large data error bars

Galaxy	Mag (band)	R_{eff}	$M/L_\star$	L	δ	β	χ^2/dof
NGC3379	− 19.8(B)	2.2	6 (7)	6	− 0.75	0.5(<0.8)	14/25
NGC4374	− 21.3(V)	3.4	6 (6)	24	− 0.88	0.01(0.01)	14/39
NGC4494	− 20.5(B)	6.1	3 (4)	20	− 0.79	0.5(0.5)	18/43

1.4 Conclusions

Modifying GR has impact at all scales so that, provided no departures from standard results, well established at Solar System scales, one cannot exclude a priori that the gravitational potential generated by a mass source has not the usual Keplerian fall off, $\Phi \propto 1/r$, but a weaker one. Here we have considered the case of a Yukawa-like correction, *i.e.* $\Phi \propto (1/r)[1 + \delta \exp(-r/L)]$ where the scale length L and the parameter δ are related to the further degrees of freedom of generalized gravity. In particular, as we have seen, this kind of potential comes out in the weak field limit of $f(R)$-gravity, the simplest extension of GR. Provided L is much larger than the Solar System scale, the corrections to the potential can significantly boost the circular velocity for extended self-gravitating systems like spiral and elliptical galaxies, galaxy clusters etc. In strong field regime, such corrections could give rise to peculiar stellar structure or trigger the Jeans instability [21]. Although a more detailed analysis is needed, we would finally stress that our analysis points towards a new approach to study the missing matter problem *without* adding dark matter components. Looking for inconsistencies rather than for agreement between these data and Newtonian models can indeed tell us not only whether the dark matter particles properties should be modified or not, but also whether our assumptions on the underlying theory of gravity are correct or not. Although it is likely that a definitive answer on this question could not be achieved in this way, the analysis of the rotation curves data, stellar dynamics and galaxy morphology stands out as a new tool to deal with modified gravity at scales complementary to those tested by cosmological probes. Asking for consistency among the results on such different scales could help us to select the correct law governing the dominant force of the universe.

References

1. J.H. Oort, Bull. Astr. Neth. **6**, 249 (1932)
2. F. Zwicky, Helv. Phys. Acta. **6**, 110 (1933)
3. S. Capozziello, V. Faraoni, *Beyond Einstein gravity…, Fund. Theories of Physics*, Vol. **170**, Springer, Heidelberg (2010)
4. C.M. Will, *Theory and Experiment in Gravitational Physics*, (2nd ed.), Cambridge University Press, Cambridge, UK (1993)
5. M. Milgrom, Astrophys. J. **270** 365 (1983)
6. T. Bernal, S. Capozziello, J.C. Hidalgo, S. Mendoza, Eur. Phys. J. **C 71**, 1794 (2011)
7. S. Capozziello, S. Tsujikawa, Phys. Rev. **D 77**, 107501 (2008)
8. S. Capozziello and A. Stabile, Class. Quant. Grav. **26**, 085019 (2009)
9. V.F. Cardone, S. Capozziello Mon. Not. R. Astron. Soc. **414**, 1301 (2011)
10. S. Capozziello, E. Filippis, V. Salzano, Mon. Not. Roy. Astron. Soc. **394**, 947 (2009)
11. R. H. Sanders, A&A **136**, L21 (1984)
12. S. Capozziello, V.F. Cardone, A. Troisi, Mon. Not. Roy. Astron. Soc. **375**, 1423 (2007)
13. V.F. Cardone, M. Sereno, A&A **438**, 545 (2005)
14. M.J illiams et al. AIP Conf. Proc., **1240**, 431 (2010) (preprint arXiv:0912.5088)
15. N.R. Napolitano, M. Capaccioli, et al., A.J. Romanowsky, Mon. Not. Roy. Astron. Soc. **357**, 691 (2005)
16. J.S. Bullock, T.S. Kolatt et al., Y. Sigad, Mon. Not. Roy. Astron. Soc. **321**, 559 (2001)
17. N.R. Napolitano, S. Capozziello, A.J. Romanowsky, M. Capaccioli and C. Tortora, ApJ **747**, 1 (2012)
18. P. Kroupa, Mon. Not. Roy. Astron. Soc. **322**, 231 (2001)
19. E.L. Lokas, Mon. Not. Roy. Astron. Soc. **333**, 697 (2002)
20. J.F. Navarro, C. S. Frenk, S.D.M. White, ApJ **490**, 493 (1997)
21. S. Capozziello, M. De Laurentis, I. De Martino, M. Formisano, S.D. Odintsov, Phys. Rev. **D 85**, 044022 (2012)

Chapter 2
Nine Years of *f(R)* Gravity and Cosmology

Valerio Faraoni

Abstract $f(R)$ gravity was reintroduced in cosmology as an alternative to dark energy in explaining the current acceleration of the universe. This area of research is reviewed with emphasis on the theoretical viability of $f(R)$ gravity, stability, weak-field limit, and recent developments, as well as some recent work on scalar-tensor and $f(R)$ black holes which are the endpoint of gravitational collapse.

2.1 Introduction

$f(R)$ gravity was introduced already in the early days of General Relativity ("GR") to explore possible alternatives [1, 2] and was studied sporadically by a few authors [3] until attempts to renormalize GR [4] showed the need of higher order corrections to the Einstein-Hilbert action, while approaches to quantum gravity introduced higher order terms and scalar fields coupling non-minimally with the spacetime curvature [5]. With the 1998 discovery of the acceleration of the universe using type Ia supernovae [6], dark energy was introduced in the standard cosmological model to account for 73% of the energy content of the universe [7]. It is difficult to conceive of a cosmological constant Λ whose energy density is fine-tuned by $\sim$ 120 orders of magnitude to account for the present acceleration. Attempts to explain the cosmic acceleration in the context of GR and without dark energy, using the backreaction of inhomogeneities on the cosmic dynamics [8] or by postulating that we live near the centre of a giant void in a dust-dominated universe [9], have so far been unconvincing. An alternative is to abandon GR and modify the Einstein-Hilbert action by replacing the Ricci scalar R with a non-linear function $f(R)$ [10, 11],

$$
\begin{aligned}
S_{EH} &= \frac{1}{2\kappa}\int d^4x\sqrt{-g}\,R + S^{(m)} \\
&\longrightarrow S = \frac{1}{2\kappa}\int d^4x\sqrt{-g}\,f(R) + S^{(m)}, \qquad (2.1)
\end{aligned}
$$

V. Faraoni (✉)
Physics Department and STAR Research Cluster, Bishop's University,
J1M 1Z7, Sherbrooke, Canada
e-mail: vfaraoni@ubishops.ca

C. Moreno González et al. (eds.), *Accelerated Cosmic Expansion*,
Astrophysics and Space Science Proceedings 38,
DOI 10.1007/978-3-319-02063-1_2,

where $\kappa \equiv 8\pi G$ and $S^{(m)}$ is the matter part of the action. Quantum corrections to the Einstein-Hilbert Lagrangian of the type $f(R) = R + \alpha R^2$ were already present in the first inflationary model of the early universe at large curvatures [12] and extra motivation for the $f(R)$ theories used more recently in cosmology comes from stringy physics [13] and from the asymptotic safety scenario [14].

The first model of $f(R)$ gravity attempting to explain the present-day cosmic acceleration was of the form $f(R) = R - \mu^4/R$ with a mass scale $\mu \sim H_0^{-1} \approx 10^{-33}$ eV [10–15]. This particular model was soon ruled out because of a catastrophic instability [16] and because it violates the post-Newtonian tests of GR [17], but the $f(R)$ proposal flourished. In principle one could include in the action other curvature invariants, as in $f\left(R, R_{ab}R^{ab}, R_{abcd}R^{abcd}, \ldots\right)$. However, the theory then generally contains ghosts [18–20]. In principle, the metric tensor g_{ab} contains six degrees of fredom which appear when the Einstein-Hilbert Lagrangian is modified to $\sqrt{-g}\, f(R, R_{ab}R^{ab}, R_{abcd}R^{abcd}, \ldots)$. In $f(R)$ gravity we will be concerned only with the usual massless spin 2 modes of GR, plus a massive scalar mode. The $f(R)$ proposal for cosmology has been taken rather seriously by the cosmological community, judging from the approximately 1100 papers which appeared on this subject since 2003 (see [21–24], for reviews and [25] for short introductions). Although the activity in this area seems to have peaked, this is still an active area of research. In this talk I will consider $f(R)$ gravity as a proof of principle that modifying gravity can explain the cosmic acceleration without dark energy and as a relatively simple but interesting alternative to GR keeping in mind that, in the road to quantization or to scenarios in which GR emerges from more fundamental constituents, GR will certainly fail somewhere, and that accompanying *infrared* corrections to the Einstein-Hilbert action are expected.

2.2 Versions of *f(R)* Gravity

Many works in the literature focus on specific choices of the function $f(R)$ but even if the $f(R)$ proposal were ultimately the correct answer to the puzzle of the cosmic acceleration (which we do not claim), there is no indication on the form of the function $f(R)$, except for the fact that it must be very close to $f(R) = R$ at Solar System scales. Therefore, we will consider only the general features of $f(R)$ gravity. There are three versions of $f(R)$ modified gravity: metric (or second order) formalism, Palatini (or first order) formalism, and metric-affine gravity.

2.2.1 *Metric f(R) Gravity*

In the metric formalism [10, 11] one varies the action

$$S = \frac{1}{2\kappa} \int d^4x \sqrt{-g}\, f(R) + S^{(m)} \tag{2.2}$$

only with respect to the (inverse) metric g^{ab}, i.e., the connection is the metric one. The resulting field equations are

$$f'(R)R_{ab} - \frac{f(R)}{2} g_{ab} = \nabla_a \nabla_b f'(R) - g_{ab} \Box f'(R) + \kappa\, T_{ab}, \tag{2.3}$$

where a prime denotes differentiation with respect to R, $\Box = g^{ab}\nabla_a \nabla_b$, and ∇_a is the covariant derivative of g_{ab}. The field equations are clearly of fourth order. Tracing eq. (2.3) yields

$$3\Box f'(R) + Rf'(R) - 2f(R) = \kappa\, T, \tag{2.4}$$

which makes it clear that $f'(R)$ is a propagating degree of freedom (by contrast, the trace equation of GR is simply the algebraic equation $R = -\kappa\, T$). Eq. (2.3) can be rewritten as the effective Einstein equation

$$G_{ab} = \frac{\kappa}{f'} \left(T_{ab} + T_{ab}^{(eff)} \right), \tag{2.5}$$

where

$$T_{ab}^{(eff)} = \frac{1}{\kappa} \left[\frac{f(R) - Rf'(R)}{2} g_{ab} + \nabla_a \nabla_b f'(R) - g_{ab} \Box f'(R) \cdot \right]. \tag{2.6}$$

The positivity of the effective gravitational coupling $G_{eff} \equiv G/f'(R)$ requires $f'(R) > 0$ (the graviton carries positive kinetic energy).

In a spatially flat Friedmann-Lemaître-Robertson-Walker (FLRW) universe with line element $ds^2 = -dt^2 + a^2(t)\left(dx^2 + dy^2 + dz^2\right)$, the field equations become

$$H^2 = \frac{1}{3f'(R)} \left[\kappa \rho^{(m)} + \frac{Rf'(R) - f(R)}{2} - 3\, H\dot{R} f''(R) \right], \tag{2.7}$$

$$2\dot{H} + 3\, H^2 = -\frac{1}{f'(R)} \left[\kappa\, P^{(m)} + f'''(R) \left(\dot{R}\right)^2 + 2\, H\dot{R} f''(R) + \ddot{R} f''(R) + \frac{f(R) - Rf'(R)}{2} \right], \tag{2.8}$$

where an overdot denotes differentiation with respect to t and $H \equiv \dot{a}/a$ is the Hubble parameter. The corresponding phase space is a 2-dimensional rather complicated subset of a 3-dimensional space [26].

2.2.2 Palatini f(R) Gravity

The Palatini (or first order) formalism of $f(R)$ gravity was also introduced to explain the cosmic acceleration without dark energy [15] The action is

$$S_{Palatini} = \frac{1}{2\kappa} \int d^4x \sqrt{-g}\, f(\tilde{R}) + S^{(m)} \left[g_{ab}, \psi^{(m)} \right]. \tag{2.9}$$

There are two Ricci tensors: R_{ab}, which is constructed with the metric connection of g_{ab}; and $\tilde{R}_{ab} = \tilde{R}_{ab}[\Gamma^{\rho}_{\mu\nu}]$, constructed with the non-metric connection $\Gamma^{\rho}_{\mu\nu}$ (and with contraction $\tilde{R} = g^{ab}\tilde{R}_{ab}$). That is, it is not assumed that the connection is the metric connection: the metric g_{ab} and the connection $\Gamma^{\alpha}_{\mu\nu}$ are treated as independent variables. When $f(R)$ is linear (the case of GR with a cosmological constant), this point of view is inconsequential and the Palatini variation produces the same field equations (the Einstein equations) as the metric variation. However, when $f(R)$ is non-linear, the variation with respect to g^{ab} yields

$$f'(\tilde{R})\tilde{R}_{ab} - \frac{f(\tilde{R})}{2} g_{ab} = \kappa\, T_{ab}, \tag{2.10}$$

which are second order equations (there are no second derivatives $\nabla_a\nabla_b f', \Box f'$). The variation with respect to the (non-metric) connection yields

$$\frac{1}{\sqrt{-g}}\, \tilde{\nabla}_d \left(\sqrt{-g}\, f'(\tilde{R}) g^{ab}\right) - \tilde{\nabla}_d \left(\sqrt{-g}\, f'(\tilde{R}) g^{d(a}\right) \delta^{b)}_c = 0, \tag{2.11}$$

where $\tilde{\nabla}_a$ is the covariant derivative of the non-metric connection Γ.

Palatini $f(R)$ gravity has been shown to contain a non-dynamical scalar field $f'(\tilde{R})$ and to run into problems when building Newtonian polytropic stars [27] and with the Cauchy problem in matter [28]. Because of these problems, we will no longer consider the Palatini version of $f(R)$ gravity here.

2.2.3 *Metric-affine f(R) Gravity*

In metric-affine $f(R)$ gravity [29], not only the metric g_{ab} and the connection $\Gamma^{\rho}_{\mu\nu}$ are independent variables, but the matter part of the action is allowed to depend explicitly on this connection. The action has the form

$$S_{affine} = \frac{1}{2\kappa} \int d^4x \sqrt{-g}\, f\left(\tilde{R}\right) + S^{(m)}\left[g_{ab}, \Gamma^{\rho}_{\mu\nu}, \psi^{(m)}\right]. \tag{2.12}$$

The possibility of a non-symmetric connection and of non-vanishing torsion is rather natural in this class of theories, which may lead to a revival of the torsion theories popular in the 1970's. Work on metric-affine gravity has been quite limited [29] because of the complication of the field equations and because cosmological accelerating dynamics can be obtained with less effort in simpler theories (which should not dismiss fundamental motivation of a different nature for these theories).

There are also "hybrid" formalisms which interpolate between the metric and the Palatini versions [30]. In the following we focus exclusively on metric $f(R)$ gravity.

2.3 Equivalence with Scalar-Tensor Gravity

If $f''(R) \neq 0$, metric and Palatini $f(R)$ gravities can be recast as $\omega = 0$ and $\omega = -3/2$ Brans-Dicke theories with a special potential, respectively [31]. Therefore, these classes of theories are nothing but scalar-tensor gravity, albeit of a form which had not been studied in depth before 2003 (see [32] for reviews of scalar-tensor gravity; this fact testifies of how little was really understood in scalar-tensor gravity).

Let us see how the reduction works for metric $f(R)$ gravity. Begining with the action (2.2), consider the extra scalar field $\phi = R$ and the equivalent action

$$S = \frac{1}{2\kappa} \int d^4x \sqrt{-g} \left[\psi(\phi) R - V(\phi) \right] + S^{(m)}, \tag{2.13}$$

where $\psi(\phi) = f'(\phi)$ and

$$V(\phi) = \phi f'(\phi) - f(\phi). \tag{2.14}$$

If $\phi = R$, the action (2.13) reduces to (2.2). Vice-versa, varying (2.13) with respect to g^{ab} yields

$$R_{ab} - \frac{1}{2} g_{ab} R = \frac{1}{\psi} \left(\nabla_a \nabla_b \psi - g_{ab} \Box \psi - \frac{V}{2} g_{ab} \right) + \frac{\kappa}{\psi} T_{ab}, \tag{2.15}$$

while variation with respect to ϕ yields $R \frac{d\psi}{d\phi} - \frac{dV}{d\phi} = (R - \phi) f''(\phi) = 0$ and $\phi = R$ if $f'' \neq 0$. Then, the massive scalar $\phi = R$ or, alternatively, $\psi = f'$ is dynamical and satisfies the trace equation

$$3 \Box \psi + 2U(\psi) - \psi \frac{dU}{d\psi} = \kappa T. \tag{2.16}$$

In terms of ψ, the action is

$$S = \frac{1}{2\kappa} \int d^4x \sqrt{-g} \left[\psi R - U(\psi) \right] + S^{(m)}, \tag{2.17}$$

where the potential is defined implicitely [33] by $U(\psi) = V(\phi(\psi)) - f(\phi(\psi))$. This is the action of a Brans-Dicke theory with Brans-Dicke parameter $\omega = 0$. The condition $f'' \neq 0$ ensures that the change of variable $R \rightarrow \psi(R)$ is invertible.

2.4 Criteria for Viability

Modifying gravity is *a priori* dangerous because the successes of GR and even of Newtonian gravity could easily be spoiled. Criteria for viability include a correct cosmological dynamics, stability, the absence of ghosts, correct Newtonian and post-Newtonian limit, a well-posed Cauchy problem, and cosmological perturbations compatible with the cosmic microwave background and large-scale structure.

Correct cosmological dynamics includes an early inflationary era (or an epoch replacing inflation with equal success), followed by a radiation era (well constrained by primordial nucleosynthesis) and by a matter era during which galaxies and other structures can grow, followed by the present accelerated era. The transitions between consecutive eras must be smooth. In the past, problems were been pointed out with the exit from the radiation era in some models [34], but these have now been solved. One can prescribe an arbitrary expansion history $a(t)$ and integrate an ODE for the function $f(R)$ which produces it ("designer $f(R)$ gravity") [35]. The function $f(R)$ thus determined is not unique and, in general, does not assume a simple form of the kind encountered in the simple models.

Stability. The prototype $f(R)$ model given by the choice $f(R) = R - \mu^4/R$ suffers from the now notorious Dolgov-Kawasaki instability [16], the study of which has been generalized to arbitrary metric $f(R)$ gravity [36] and extended to more general theories [37]. Parametrize the deviations from GR as

$$f(R) = R + \epsilon\, \varphi(R), \tag{2.18}$$

where ϵ is a positive smallness parameter and $f''(R) \neq 0$. Then the trace equation gives

$$\Box R + \frac{\varphi'''}{\varphi''} \nabla^c R \nabla_c R + \left(\frac{\epsilon\varphi' - 1}{3\epsilon\varphi''}\right) R = \frac{\kappa\, T}{3\epsilon\varphi''} + \frac{2\varphi}{3\varphi''}. \tag{2.19}$$

Expanding locally the metric and the Ricci scalar as $g_{ab} = \eta_{ab} + h_{ab}$ and $R = -\kappa\, T + R_1$ (where η_{ab} is the Minkoswki metric and h_{ab} and R_1 are perturbations) one obtains, to first order

$$\begin{aligned} &\ddot{R}_1 - \nabla^2 R_1 - \frac{2\kappa\varphi'''}{\varphi''} \dot{T}\dot{R}_1 + \frac{2\kappa\varphi'''}{\varphi''} \vec{\nabla} T \cdot \vec{\nabla} R_1 \\ &+ \frac{1}{3\varphi''}\left(\frac{1}{\epsilon} - \varphi'\right) R_1 = \kappa\, \ddot{T} - \kappa\nabla^2\, T - \frac{\left(\kappa T\varphi^2 + 2\varphi\right)}{3\varphi''}. \end{aligned} \tag{2.20}$$

The coefficient of R_1 yields the effective mass squared $m^2 \simeq \frac{1}{3\epsilon\varphi''}$ and, therefore, the scalar degree of freedom is stable if $f''(R) > 0$ and unstable if $f''(R) < 0$. For example, the prototype model $f(R) = R - \mu^4/R$ has $f'' < 0$ and the instability manifests itself on the time scale (dictated by the value of $\mu \simeq H_0^{-1}$) $t \sim 10^{-26}$ s [16]. A physical interpretation of this stability criterion is the following [38]: the effective gravitational coupling is $G_{eff} = G/f'(R) > 0$. If $dG_{eff}/dR = -f''G/(f')^2 > 0$, G_{eff} increases with R and, at large curvatures, causes gravity to become stronger, which causes an even larger R, in a positive feedback. *Vice-versa*, if $dG_{eff}/dR < 0$ then G_{eff} does not increase as the curvature increases and there is stability.

The previous stability analysis assumes a flat background and, therefore, is only valid in the limit of small wavelengths. A stability analysis can be performed analytically for de Sitter space (which is often a late-time attractor of the cosmological dynamics). A study with a covariant and gauge-invariant formalism (necessary for long-wavelength cosmological perturbations) [39] yields the first order stability criterion

$$\frac{(f_0')^2 - 2f_0 f_0''}{f_0' f_0''} \geq 0, \tag{2.21}$$

where the zero subscript describes the fact that quantities are evaluated in the background de Sitter space of constant curvature R_0.

Beyond the linear approximation, certain metric $f(R)$ models have been found to suffer from a non-linear instability, which makes it difficult to construct relativistic stars in the presence of strong gravity because of a singularity developing at large curvature [40]. However, the problem may be cured by adding a small quadratic term to $f(R)$ [41] and it appears to be related to the complication of working with an implicit potential which becomes ill-defined in the scalar-tensor description of metric $f(R)$ gravity [42].

Weak-field limit. The weak-field limit of metric $f(R)$ gravity was studied in [43–45], following early work on particular models [21, 22]. The idea is to compute the PPN parameter γ through the PPN expansion of the line element

$$\begin{aligned} ds^2 =& -\left[1 + 2\Psi(r) - H_0^2 r^2\right] dt^2 \\ &+ \left[1 + 2\Phi(r) + H_0^2 r^2\right] dr^2 + r^2 \left(d\theta^2 + \sin^2\theta \, d\varphi^2\right) \end{aligned} \tag{2.22}$$

in Schwarzschild coordinates, where $|\Psi(r)|, |\Phi(r)| \ll 1$ and $H_0 r \ll 1$, $R(r) = R_0 + R_1$. The PPN parameter γ is given by $\gamma = -\Phi(r)/\Psi(r)$ [46]. Under the assumptions that [43] $f(R)$ is analytical at R_0, $mr \ll 1$ (where m is the effective mass of the scalar degree of freedom), and that the matter composing the spherical body satisfies $P \simeq 0$, $T = T_0 + T_1 \simeq -\rho$, the trace equation yields

$$\nabla^2 R_1 - m^2 R_1 = \frac{-\kappa\rho}{3f_0''}, \tag{2.23}$$

where

$$m^2 = \frac{(f_0')^2 - 2f_0 f_0''}{3f_0' f_0''} \tag{2.24}$$

(the same expression obtained with gauge-invariant analysis of de Sitter space or by calculating the propagator). If $mr \ll 1$, the solution is

$$\Psi(r) = \frac{-\kappa M}{6\pi f_0'} \frac{1}{r}, \qquad \Phi(r) = \frac{\kappa M}{12\pi f_0'} \frac{1}{r}, \tag{2.25}$$

from which one obtains the PPN parameter $\gamma = -\Phi(r)/\Psi(r) = 1/2$ in violent constrast with the Cassini limit $|\gamma - 1| < 2.3 \cdot 10^{-5}$ [47]. This result would be the end of metric $f(R)$ gravity if it wasn't for the fact that mr is not always small. Due to the *chameleon effect*, the range of the scalar field degree of freedom and its effective mass m depend on the curvature and the energy density of the environment. This range λ can be small ($m > 10^{-3}$ eV and $\lambda < 0.2$ mm) at Solar System densities and large (of the order of the Hubble radius H_0^{-1}) at cosmological densities [20, 48]. The chameleon effect is not imposed by hand but is built into $f(R)$ gravity models. It is analogous to the chameleon mechanism of quintessence models with potentials $V(\phi) \approx 1/\phi^\alpha$ (with $\alpha > 0$) [49]. Several explicit forms of the function $f(R)$ are now known to exhibit an efficient chameleon mechanism sheltering the Solar System tests from modified gravity, which shows its effects at large scales [21–23].

Cosmological perturbations Most theories of gravity alternative to GR admit FLRW solutions [23], therefore, obtaining the correct dynamics for a FLRW universe does not discriminate between competing theories of gravity and cosmology. However, the growth history of non-linear cosmological perturbations is sensitive to the theory of gravity and its study has the potential to discriminate between different models in the near future. Recently, work on $f(R)$ gravity has moved to the study of perturbations and its comparison with large-scale structure surveys.

By assuming a cosmic evolution $a(t)$ typical of the ΛCDM model it is found (*e.g.*, [50]) that, to lowest order, vector and tensor modes are unaffected by $f(R)$ corrections to the Einstein-Hilbert action. The condition $f''(R) \geq 0$ for the stability of scalar modes is recovered in this context, and $f(R)$ corrections do affect the scalar modes. Large angle anisotropies in the cosmic microwave background are produced, as well as different correlations (compared with the case of Einstein graity) between cosmic microwave background and galaxy surveys.

The Cauchy problem A well-posed initial value problem is necessary in order for a physical theory to have predictive power. The Cauchy problem for metric and Palatini $f(R)$ gravity was studied in [28] (see also the early work of [51, 52]). Using the equivalence between $f(R)$ and scalar-tensor gravity, the Cauchy problem was reduced to the analogous one for Brans-Dicke gravity and, taking advantage of previous results for scalar-tensor gravity [53], the following was established. The initial value problem for metric $f(R)$ gravity is well-posed *in vacuo* and for "reasonable" forms of matter (meaning, the same for which the Cauchy problem of GR is well-posed [54]). The Cauchy problem is also well-posed for Palatini $f(R)$ gravity *in vacuo* (in which case the theory reduces to GR with a cosmological constant), but is extremely unlikely to be well-posed or well-formulated in matter, a problem that is reminiscent of the difficulties in constructing matter configurations [27] and is ultimately due to the structure of the field equations—this adds to the problems of Palatini $f(R)$ gravity [21, 55, 56] (but see [57] for a different view).

2.5 Scalar-Tensor and *f(R)* Black Holes

In any theory of gravity it is important to understand the structure of its spherically symmetric solutions and, in particular, of its black holes. In GR stationary black holes, which are the endpoint of gravitational collapse, must be axisymmetric [58].

In scalar-tensor (and, therefore, in $f(R)$ gravity) Birkhoff's theorem is lost and spherical, asymptotically flat, black holes are not forced to be static. Let us consider Brans-Dicke theory in the Jordan frame, described by the action

$$S_{BD} = \int d^4x\sqrt{-\hat{g}}\left[\varphi\hat{R} - \frac{\omega_0}{\varphi}\hat{\nabla}^{\mu}\varphi\hat{\nabla}_{\mu}\varphi + L_m(\hat{g}_{\mu\nu}, \psi)\right] \tag{2.26}$$

A 1972 theorem by Hawking [58] states that the endpoint of axisymmetric collapse in this theory must be a GR black hole. This result was generalized (for spherical symmetry only) in [59]. A simple proof for more general stationary spacetimes became available recently [60], which extends Hawking's result to the general scalar-tensor theory described by the action

$$S_{ST} = \int d^4x\sqrt{-\hat{g}}\left[\varphi\hat{R} - \frac{\omega(\varphi)}{\varphi}\hat{\nabla}^{\mu}\varphi\hat{\nabla}_{\mu}\varphi - V(\varphi) + L_m(\hat{g}_{\mu\nu}, \psi)\right]. \tag{2.27}$$

We assume 1) asymptotic flatness (realistic astrophysical collapse occurs on scales much smaller than H_0^{-1}), therefore the Brans-Dicke scalar field $\varphi \to \varphi_0$ as $r \to +\infty$, $V(\varphi_0) = 0$, and $\varphi_0\, V'(\varphi_0) = 2\, V(\varphi_0)$. 2) Stationarity (the black hole is the endpoint of gravitational collapse).

Using the Einstein frame variables $\hat{g}_{\mu\nu} \to g_{\mu\nu} = \varphi\, \hat{g}_{\mu\nu}$, $\varphi \to \phi$ with $d\phi = \sqrt{\frac{2\omega(\varphi)+3}{16\pi}}\,\frac{d\varphi}{\varphi}$ $(\omega \neq -3/2)$, the action is cast into the form

$$S_{ST} = \int d^4x\sqrt{-g}\left[\frac{R}{16\pi} - \frac{1}{2}\nabla^{\mu}\phi\nabla_{\mu}\phi - U(\phi) + L_m(\hat{g}_{\mu\nu}, \psi)\right] \tag{2.28}$$

where $U(\phi) = V(\varphi)/\varphi^2$. The Einstein frame field equations are

$$\hat{R}_{\mu\nu} - \frac{1}{2}\hat{R}\hat{g}_{\mu\nu} = \frac{\omega(\varphi)}{\varphi^2}\left(\hat{\nabla}_{\mu}\varphi\hat{\nabla}_{\nu}\varphi - \frac{1}{2}\hat{g}_{\mu\nu}\,\hat{\nabla}^{\lambda}\varphi\hat{\nabla}_{\lambda}\varphi\right) + \frac{1}{\varphi}\left(\hat{\nabla}_{\mu}\hat{\nabla}_{\nu}\varphi - \hat{g}_{\mu\nu}\hat{\Box}\varphi\right) - \frac{V(\varphi)}{2\varphi}\hat{g}_{\mu\nu}, \tag{2.29}$$

$$(2\omega + 3)\,\hat{\Box}\varphi = -\omega'\,\hat{\nabla}^{\lambda}\varphi\hat{\nabla}_{\lambda}\varphi + \varphi\, V' - 2V. \tag{2.30}$$

Since the conformal factor depends only on φ, the Einstein frame symmetries are the same as in the Jordan frame and there exists a timelike Killing vector ξ^{μ} describing stationarity and a spacelike Killing vector ζ^{μ} which describes axial symmetry. Consider, *in vacuo*, a 4-volume $\mathcal{V}$ bounded by the horizon $\mathcal{H}$, two Cauchy hypersurfaces $\mathcal{S}_1, \mathcal{S}_2$, and a timelike 3-surface at infinity. Multiply the Einstein frame field equation $\Box\phi = U'(\phi)$ by U' and integrate the result over $\mathcal{V}$. This yields

$$\int_{\mathcal{V}} d^4x\sqrt{-g}\, U'(\phi)\Box\phi = \int_{\mathcal{V}} d^4x\sqrt{-g}\, U'^2(\phi), \tag{2.31}$$

which can be rewritten as

$$\int_{\mathcal{V}} d^4x \sqrt{-g}\,[U''(\phi)\nabla^{\mu}\phi\nabla_{\mu}\phi + U'^2(\phi)]$$
$$= \int_{\partial\mathcal{V}} d^3x \sqrt{|h|}\, U'(\phi) n^{\mu}\nabla_{\mu}\phi \tag{2.32}$$

where n^{μ} is the normal to the boundary and h is the determinant of the induced metric $h_{\mu\nu}$ on this boundary. Now split the boundary into its constituents, $\int_{\mathcal{V}} = \int_{\mathcal{S}_1} + \int_{\mathcal{S}_2} + \int_{horizon} + \int_{r=\infty}$. We have $\int_{\mathcal{S}_1} = -\int_{\mathcal{S}_2}$, $\int_{r=\infty} = 0$, and

$$\int_{horizon} d^3x \sqrt{|h|}\, U'(\phi) n^{\mu}\nabla_{\mu}\phi = 0 \tag{2.33}$$

because of the symmetries. As a result, it is

$$\int_{\mathcal{V}} d^4x \sqrt{-g}\,[U''(\phi)\nabla^{\mu}\phi\nabla_{\mu}\phi + U'^2(\phi)] = 0. \tag{2.34}$$

Since $U'^2 \geq 0$, $\nabla^{\mu}\phi$ (orthogonal to both ξ^{μ}, ζ^{μ} on H) is spacelike or zero, and $U''(\phi)$ must be non-negative for stability (the black hole is the endpoint of collapse!), then it must be $\nabla_{\mu}\phi \equiv 0$ in $\mathcal{V}$ and $U'(\phi_0) = 0$. Since for ϕ =const., the theory reduces to GR, black holes must be of the Kerr type.

Now, metric $f(R)$ gravity is a special Brans-Dicke theory with $\omega = 0$ and $V \neq 0$ and is covered by the previous discussion. For $\omega = -3/2$, the vacuum theory reduces to GR and Hawking's theorem applies (Palatini $f(R)$ gravity is a special Brans-Dicke theory with $\omega = -3/2$ and $V \neq 0$). Exceptions not covered by our proof include theories in which $\omega \to \infty$ somewhere, and theories in which φ diverges at infinity or on the horizon (an example is the maverick solution of [61], which is however unstable). The proof is extended immediately to the case of electrovacuum or any conformal matter with energy-momentum trace $T = 0$ because the Brans-Dicke scalar couples only to the trace T and the equations used in the proof do not change by including such forms of matter.

Even though Birkhoff's theorem is lost, black holes which are the endpoint of axisymmetric gravitational collapse and are asymptotically flat in *general* scalar-tensor gravity are the same as in GR (*i.e.*, Kerr-Newman black holes). The exceptions are unphysical or unstable solutions which cannot be the endpoint of collapse, or do not satisfy the weak/null energy condition. Asymptotic flatness is a technical requirement, but realistic gravitational collapse in an astrophysical context occurs on scales which are much smaller than the Hubble time and the influence of an asymptotically FLRW structure during this short timescale is completely negligible. Exceptions are situations involving the formation of primordial black holes in the very early universe when the collapse and the Hubble time scales are comparable.

2.6 Conclusions

Our brief exposition does not make justice to the amount of work performed on $f(R)$ gravity and we recommend reading the reviews on this subject [21–24] and the original references. Not everything is understood about $f(R)$ gravity and some work remains to be done. What has been shown in the past nine years is that metric $f(R)$ gravity is in principle capable of reproducing the observed acceleration of the cosmic expansion while respecting basic requirements for theoretical viability and fitting the observational data. Therefore, modifying gravity (although perhaps not in the $f(R)$ way) seems feasible and deviations from GR are expected soon or later. The popularity of this subject shows that contemplating modifications of GR in relation to real observations is no longer a taboo and indeed many effective actions or corrections to GR occupy theoretical physicists today [23]. Another lesson is that much new physics can sometimes be learned from theories that have been around for a long time and were naively believed to be well understood.

Acknowledgments I thank the organizers for the invitation to this conference and the Natural Sciences and Engineering Research Council of Canada and Bishop's University for financial support.

References

1. A.S. Eddington, *The Mathematical Theory of Relativity*. (Cambridge University Press, Cambridge, 1923).
2. See H.-J. Schmidt, Int. J. Geom. Meth. Phys. **4**, (2007) 209 for an historical overview.
3. R. Bach, Math. Zeischfr. **9**, 110 (1921); C. Lanczos, Zeit. Phys. **73**, 147 (1932); E. *Schrödinger, Space-Time Structure* (Cambridge University Press, Cambridge, 1950); H. Buchdhal, Quart. J. Math. Oxford **19** 150 (1948); Proc. Nat. Acad. Sci. USA **34** 66 (1948); Acta Mathematica **85** (1951) 63; J. London Math. Soc. **26** 139 (1951); **26** 150 (1951); Proc. Edimburgh Math. Soc. **10** (1953) 16; Nuovo Cimento **23** (1962) 141; Tensor **21** (1970) 340; Proc. Cambr. Phil. Soc. **68** (1970) 179; Mon. Not. R. Astr. Soc. **150** (1970) 1; Proc. Cambr. Phil. Soc. **74** (1973) 145; J. Phys. A **11** (1978) 871; Int. J. Theor. Phys. **17** (1978) 149.
4. R. Utiyama, B. DeWitt, J. Math. Phys **3**, 608 (1962); A. Strominger, Phys. Rev. D. **30**, 2257 (1984)
5. I.L. Buchbinder, S.D. Odintsov, I.L. Shapiro, Effective Action in Quantum Gravity (IOP, Bristol, 1992); G. Vilkovisky, Class. Quantum Grav. **9**, 985 (1992); M.B. Green, G.H. Schwarz, E. Witten, *Superstring Theory* (Cambridge University Press, Cambridge, 1987)
6. A.G. Riess et al., Astron. J. **116**, 1009 (1998); S. Perlmutter et al., Nature **391**, 51 (1998); A.G. Riess et al., Astron. J. **118**, 2668 (1999); S. Perlmutter et al., Astrophys. J. **517**, 565 (1999); A.G. Riess et al., Astrophys. J. **560**, 49 (2001); J.L. Tonry et al., Astrophys. J. **594**, 1 (2003); R. Knop et al., Astrophys. J. **598**, 102 (2003); A.G. Riess et al., Astrophys. J. **607**, 665 (2004); B. Barris et al., Astrophys. J. **602**, 571 (2004)
7. E. Komatsu et al., Astrophys. J. (Suppl.) **192**, 18 (2011)
8. T. Buchert, Gen. Rel. Gravit. **32**, 105 (2000); T. Buchert, M. Carfora, Class. Quantum Grav. **19**, 6109 (2002); **25**, 195001 (2008); S. Räsänen, JCAP. **02**, 003 (2004); D.L. Wiltshire, New J. Phys. **9**, 377 (2007); Phys. Rev. Lett. **99**, 251101 (2007); E.W. Kolb, S. Matarrese, A. Riotto, New J. Phys. **8**, 322 (2006); J. Larena, T. Buchert, J.-M. Alimi, Class. Quantum Grav. **23**, 6379 (2006); A. Paranjape, T.P. Singh, Phys. Rev. D. **76**, 044006 (2007); N. Li, D.J. Schwarz, Phys.

Rev. D. **76**, 083011 (2007); **78**, 083531 (2008); T. Buchert, Gen. Rel. Gravit. **40**, 467 (2008); AIP Conf. Proc. **467**, 361 (2007); J. Larena, J.-M. Alimi, T. Buchert, M. Kunz, P. Corasaniti, Phys. Rev. D. **79**, 083011 (2008)
9. H. Alnes, M. Amarzguioui, Grn, Phys. Rev. D. **73**, 083519 (2006); D.J.H. Chung, A.E. Romano, Phys. Rev. D. **74**, 103507 (2006); K. Enqvist, T. Mattsson, JCAP **2**, 19 (2007); J. Garcia-Bellido, T. HaugbÃÿ lle, JCAP. **4**, 3 (2008); **9**, 16 (2008); S. Alexander, T. Biswas, A. Notari, D. Vaid, JCAP. **9**, 25 (2009); M.-N. Célérier, K. Bolejko, and A. Krasínski, Astron. Astrophys. **518** (2010) A21; T. Biswas, A. Notari, and W. Valkenburg, JCAP. **11** (2010) 30; V. Marra and M. Pääkkönen, JCAP. **12** (2010) 21; S. February, J. Larena, M. Smith, and C. Clarkson, Mon. Not. Roy. Astr. Soc. **405** (2010) 2231; C. Clarkson and R. Maartens, 2010, Class. Quantum Grav. **27** (2010) 124008; K. Bolejko, M.-N. Célérier and A. Krasínski, Class. Quantum Grav. **28** (2011) 164002; K. Bolejko and R.A. Sussman, Phys. Lett. B **697** (2011) 265.
10. S. Capozziello, S. Carloni, A. Troisi, arXiv:astro-ph/0303041.
11. S.M. Carroll, V. Duvvuri, M. Trodden, M.S. Turner, Phys. Rev. D **70**, 043528 (2004)
12. A.A. Starobinsky, Phys. Lett. B **91**, 99 (1980)
13. P. Candelas, G.T. Horowitz, A. Strominger, E. Witten, Nucl. Phys. B **258**, 46 (1985); S. Nojiri, S.D. Odintsov, Phys. Lett. B **576**, 5 (2003); S. Nojiri, S.D. Odintsov, and M. Sami, Phys. Rev. D. **74**, 046004 (2006)
14. M. Hindmarsh, I.D. Saltas, arXiv:1203.3957.
15. D.N. Vollick, Phys. Rev. D. **68**, 063510 (2003)
16. A.D. Dolgov, M. Kawasaki, Phys. Lett. B **573**, 1 (2003); S. Nojiri, S.D. Odintsov, Phys. Rev. D. **68**, 123512 (2003)
17. T. Chiba, Phys. Lett. B. **575**, 1 (2003)
18. K.S. Stelle, Phys. Rev. D. **16**, 953 (1977)
19. A. Nunez, S. Solganik, arXiv:hep-th/0403159; D. Comelli, Phys. Rev. D **72**, 064018 (2005)
20. I. Navarro, K. Van Acoleyen, JCAP, **0603**, 008 (2006); R. Woodard, astro-ph/0601672; A. De Felice, M. Hindmarsh, and M. Trodden, JCAP **08**, 005 (2006); G. Calcagni, B. de Carlos, and A. De Felice, Nucl. Phys. B **752**, 404 (2006)
21. T.P. Sotiriou, V. Faraoni, Rev. Mod. Phys. **82** (2010) 451.
22. A. De Felice, S. Tsujikawa, Living Rev. Rel. **13**, 3 (2010)
23. T. Clifton, P.G, Ferreira, A. Padilla, C. Skordis, *Phys. Reps.* **513** 1 (2012)
24. S. Capozziello, V. Faraoni, *Beyond Einstein Gravity* (Springer, New York, 2010).
25. N. Straumann, arXiv:0809.5148; V. Faraoni, arXiv:0810.2602; S. Capozziello, M. Francaviglia, Gen. Rel. Grav. **40**, (2008) 357; S. Nojiri, S.D. Odintsov, Int. J. Geom. Meth. Mod. Phys.. **4**,115 (2007); T.P. Sotiriou, J. Phys. Conf. Ser. **189**, 012039 (2009); S. Capozziello, M. De Laurentis, V. Faraoni, arXiv:0909.4672.
26. J.C.C. de Souza and V. Faraoni, Class. Quant. Grav. **24**, 3637 (2007); V. Faraoni, Ann. Phys. (NY) **317**, 366 (2005)
27. E. Barausse, T.P. Sotiriou, J. Miller, Class. Quantum Grav. **25**, 062001 (2008); **25**, 105008 (2008); EAS Publ. Ser. **30**, 189 (2008); B. Li, D.F. Mota, D.J. Shaw, Phys. Rev. D **78**, 064018 (2008); E.E. Flanagan, Phys. Rev. Lett. **92**, 071101 (2004)
28. N. Lanahan-Tremblay, V. Faraoni, Class. Quantum Grav. **24**, 5667 (2007); V. Faraoni and N. Lanahan-Tremblay, Phys. Rev. D **78**, 064017 (2008)
29. T.P. Sotiriou, Class. Quantum Grav. **23**, 5117 (2006); Proceedings of the Eleventh Marcel Grossmann Meeting on General Relativity, H. Kleinert, R.T. Jantzen, R. Ruffini eds., p. 1223, (World Scientific, Singapore, 2008) [arXiv:gr-qc/0611158]; arXiv:0710.4438; T.P. Sotiriou, S. Liberati, Ann. Phys. (NY) **322**, 935 (2007); J. Phys. Conf. Ser. **68**, 012022 (2007)
30. T. Koivisto, Phys. Rev. D. **83**, 101501 (2011)
31. P.W. Higgs, Nuovo Cimento **11**, 816 (1959); B. Whitt, Phys. Lett. B **145**, 176 (1984); P. Teyssandier and P. Tourrenc, J. Math. Phys. **24**, 2793 (1983); J.D. Barrow, S. Cotsakis, Phys. Lett. B **214** (1988) 515; K. Maeda, Phys. Rev. D. **39**, 3159 (1989); T. Chiba, Phys. Lett. B **575**, 1 (2005)
32. Y. Fujii, K. Maeda, *The Scalar-Tensor Theory of Gravitation.* (Cambridge University Press, Cambridge, 2003); V. Faraoni, *Cosmology in Scalar-Tensor Gravity.* (Kluwer Academic, Dordrecht, 2004).

33. The fact that the potential is not given explicitly may sometimes cause problems in the equivalent scalar-tensor picture of $f(R)$ gravity (34)
34. L. Amendola, D. Polarski, S. Tsujikawa, Phys. Rev. Lett. **98**, 131302 (2007); S. Capozziello, S. Nojiri, S.D. Odintsov, A. Troisi, Phys. Lett. B **639**, 135 (2006); L. Amendola, R. Gannouji, D. Polarski, S. Tsujikawa, Phys. Rev. D **75**, 083504 (2007); S. Nojiri, S.D. Odintsov, Phys. Rev. D **74**, 086005 (2006); A.W. Brookfield, C. van de Bruck, L.M.H. Hall, Phys. Rev. D **74**, 064028 (2006)
35. S. Capozziello, V. Cardone, A. Troisi, Phys. Rev. D **71**, 043503 (2005); S. Capozziello, S. Nojiri, S.D. Odintsov, A. Troisi, Phys. Lett. B **639**, 135 (2006); S. Nojiri, S.D. Odintsov, Phys. Rev. D **74**, 086005 (2006); J. Phys. A **40**, 6725 (2007); J. Phys. Conf. Ser. **66**, 012005 (2007); A. de la Cruz-Dombriz, A. Dobado, Phys. Rev. D **74**, 087501 (2006); S. Fay, S. Nesseris, L. Perivolaropoulos, Phys. Rev. D **76**, 063504 (2007)
36. V. Faraoni, Phys. Rev. D **74**, 104017 (2006); see also M.D. Seifert, Phys. Rev. D. **76**, 064002 (2007)
37. G. Cognola, S. Zerbini, J. Phys. A **39** 6245 (2006); Int. J. Theor. Phys. **47** 3186 (2008); G. Cognola, E. Elizalde, S. Nojiri, S.D. Odintsov, S. Zerbini, JCAP **0502**, 010 (2005); G. Cognola, M. Gastaldi, S. Zerbini, Int. J. Theor. Phys. **47** 898 (2008)
38. V. Faraoni, Phys. Rev. D. **75**, 067302 (2007)
39. V. Faraoni, Phys. Rev. D. **70**, 044037 (2004); Phys. Rev. D. **72**, 061501(R) (2005); V. Faraoni, S. Nadeau, Phys. Rev. D **72**, 124005 (2005)
40. A.V. Frolov, Phys. Rev. Lett **101**, 061103 (2008); S.A. Appleby, R.A. Battye, Phys. Lett. B. **654**, 7 (2007); T. Kobayashi, K. Maeda, Phys. Rev. D. **78**, 064019 (2008); Phys. Rev. D **79**, 024009 (2009); see also E. Babichev, D. Langlois, Phys. Rev. D. **80**, 121501(R) (2009); Phys. Rev. D **81**, 124051 (2010)
41. F. Briscese, E. Elizalde, S. Nojiri, S.D. Odintsov, Phys. Lett. B, **646**, 105 (2007); K. Bamba, S. Nojiri, S.D. Odintsov, JCAP, **10**, (2008) 45; A. Upadhye, W. Hu, Phys. Rev. D **80** 064002 (2009); A.A. Appleby, R.A. Battye, A.A. Starobinsky, JCAP **6**, 5 (2010)
42. L.G. Jaime, L. Patiño, M. Salgado, Phys. Rev. D. **83**, 024039 (2011); arXiv:1206.1642
43. T. Chiba, T.L. Smith, A.L. Erickcek, Phys. Rev. D. **75**, 124014 (2007)
44. A. Erickcek, T.L. Smith, and M. Kamionkowski, Phys. Rev. D. **74**, 121501(R) (2006)
45. G.J. Olmo, Phys. Rev. D. **75**, 023511 (2007)
46. C.M. Will, *Theory and Experiment in Gravitational Physics, 2nd edition.* (Cambridge University Press, Cambridge, 1993).
47. B. Bertotti, L. Iess, P. Tortora, Nature **425**, 374 (2003)
48. T. Faulkner, M. Tegmark, E.F. Bunn, Y. Mao, Phys. Rev. D. **76**, 063505 (2007)
49. J. Khoury, A. Weltman, Phys. Rev. Lett. **93**,171104 (2004); Phys. Rev. D **69**, 044026 (2004)
50. Y.-S. Song, W. Hu, I. Sawicki, Phys. Rev. D **75**, 044004 (2007); S. Nojiri, S.D. Odintsov, J. Phys. Conf. Ser. **66**, (2007) 012005; W. Hu, I. Sawicki, Phys. Rev. D **76**, 104043 (2007); **76** 064004 (2007); A. de la Cruz-Dombriz, A. Dobado, Phys. Rev. D **74**, 087501 (2006); S. Fay, S. Nesseris, L. Perivolaropoulos, Phys. Rev. D **76**, 063504 (2007); T. Faulkner, M. Tegmark, E.F. Bunn, Y. Mao, Phys. Rev. D **76**, 063505 (2007); A.A. Starobinsky, JETP Lett. **86**, 157 (2007); L.M. Sokolowski, Class. Quantum Grav. **24**, 3713 (2007); S. Nojiri and S.D. Odintsov, J. Phys. A **40**, 6725 (2007)
51. W.J. Cocke, J.M. Cohen, J. Math. Phys. **9**, (1968) 971.
52. D.R. Noakes, J. Math. Phys. **24**, (1983) 1846.
53. M. Salgado, Class. Quant. Grav. **23**, 4719 (2006); M. Salgado, D. Martinez-del Rio, M. Alcubierre, D. Nunez, Phys. Rev. D **77**, 104010 (2008)
54. R.M. Wald, General Relativity (Chicago University Press, Chicago, 1984).
55. A. Iglesias, N. Kaloper, A. Padilla, M. Park, Phys. Rev. D **76**, 104001 (2007)
56. V. Faraoni, Phys. Lett. B. **665**, 135 (2008)
57. S. Capozziello, S. Vignolo, Class. Quantum Grav. **26**, 168001 (2009); S. Capozziello and S. Vignolo, Int. J. Geom. Meth. Mod. Phys. **9**, 1250006 (2012)
58. S.W. Hawking, Commun. Math. Phys. **25**, 152 (1972)

59. A.E. Mayo, J.D. Bekenstein, Phys. Rev. D **54**, 5059 (1996); J.D. Bekenstein, arXiv:gr-qc/9605059.
60. T.P. Sotiriou, V. Faraoni, Phys. Rev. Lett. **108**, 081103 (2012)
61. N. Bocharova, K. Bronnikov, V. Melnikov, Vestn. Mosk. Univ. Fiz. Astron. **6** 706 (1970); K.A. Bronnikov, Yu.N. Kireyev, Phys. Lett. A **67**, 95 (1978)

Chapter 3
A Geometrical Approach to Brans-Dicke Theory

M. L. Pucheu, T. S. Almeida and C. Romero

Abstract We consider an approach to Brans-Dicke theory of gravity in which the scalar field is considered as a geometrical field. By postulating the Palatini variation, we find out that the role played by the scalar field consists of turning the space-time geometry into a Weyl integrable manifold. This assumption leads to a scalar-tensor theory that presents new features. In the light of this theory we examine how the theory passes the classical tests in the solar system.

3.1 Introduction

Einstein's General Relativity represents certainly one of the most important intelectual developments of the XX century. Einstein based his theory on the already known results of the Newtonian gravity, the equivalence of inertial and gravitational masses, and raised this result to the status of a principle: the Einstein's Principle of Equivalence (EEP) [1]. This principle has its mathematical formulation given by the Geodesic Postulate, which establishes that free particles under the influence of gravity will follow geodesics, introducing in this way the Riemannian geometry.

On the other hand, another theory of gravitation, due to Brans and Dicke (BD), extends Einstein's theory by incorporating the Mach's Principle. This idea is materialized by including a scalar field that plays the role of the inverse of the gravitational coupling, which becomes a function of the space-time. In its original formulation, this theory satisfies the Einstein Equivalence Principle and the Geodesic Postulate, in the same way as GR does [2].

As can be found in many textbooks, there are actually two variational principles that one can apply to the Einstein-Hilbert action in order to derive de Einstein's equations: the standard metric variation and a less standard variation, namely, the Palatini variation. In the latter, the metric $g_{\mu\nu}$ and the connection $\Gamma^{\alpha}_{\mu\nu}$ are assumed to be independent variables and one varies the action with respect to both of them, under

M. L. Pucheu (✉) · T. S. Almeida · C. Romero
Departamento de Física, Universidade Federal da Paraíba,
Jõao Pessoa, Brasil
e-mail: mlaurapucheu@fisica.ufpb.br

C. Romero
e-mail: cromero@fisica.ufpb.br

C. Moreno González et al. (eds.), *Accelerated Cosmic Expansion*,
Astrophysics and Space Science Proceedings 38,
DOI 10.1007/978-3-319-02063-1_3, © Springer International Publishing Switzerland 2014

the important assumption that the matter action does not depend on the connection [3]. The Palatini formulation of GR is equivalent to the metric version of this theory as a consequence of the fact that the variation with respect to the connection $\Gamma^{\alpha}_{\mu\nu}$ gives the Levi-Civita connection of the metric $g_{\mu\nu}$. So, there is no particular reason to impose the Palatini variational principle in GR instead of the metric variational principle. However, this is no longer true for more general actions that depends on functions of curvature invariants [4].

In the present work, we begin by applying the Palatini variational method to the Brans-Dicke theory of gravity. The field equation corresponding to the variation of the connection leads to the compatibility condition associated to Weyl integrable geometry, where the BD scalar field is interpreted as a function of the the Weyl scalar field. Assuming that the dynamics of gravitation occurs in a space-time being described by a Weyl integrable geometry, the original coupling between matter and gravitation proposed by BD must be modified to assure that the EP is satisfied. It turns out that the new coupling between the scalar field and the matter does not violate the EEP since particles still follow geodesics, namely, Weyl geodesics.

Although Brans-Dicke theory seems to be an exhausted issue, it is worth to stress at this point that our interest lies on the geometrical ground that Palatini variational method can provide. As mentioned above, considering the metric and the connection as independent fields amounts to decoupling the metric structure of the space-time and its geodesic structure with the connection $\Gamma^{\alpha}_{\mu\nu}$ being distinct from the Levi-Civita connection. This leads to a new scenario for the usual Brans-Dicke theory, in which the scalar field has a geometrical meaning and plays a fundamental role in the movement of particles and light rays through the geodesic lines of the space-time.

The paper is organized as follows. In Sec. 2, we derive the field equations by using what we define as the *extended Palatini variational method*, where the BD scalar field is now reinterpreted as a geometrical scalar field. Matter coupling will be also discussed in this section. In Sec. 3, we find the Bianchi identities in the Weyl integrable geometry and present the modified conservation law, and in the next, we analyze the solar system experiments in the context of the proposed formalism. We conclude with some final remarks in Sec. 6.

3.2 The Field Equations

Let us start with the Brans-Dicke gravitational action [5]

$$S_G = \int d^4x\sqrt{-g}\left(\Phi R + \frac{\omega}{\Phi}\Phi^{,\alpha}\Phi_{,\alpha}\right), \tag{3.1}$$

where $R = g^{\mu\nu}R_{\mu\nu}(\Gamma)$. Here we are adopting the definition

$$R^{\alpha}_{\mu\beta\nu} = \Gamma^{\alpha}_{\beta\mu,\nu} - \Gamma^{\alpha}_{\nu\mu,\beta} + \Gamma^{\alpha}_{\rho\nu}\Gamma^{\rho}_{\beta\mu} - \Gamma^{\alpha}_{\rho\beta}\Gamma^{\rho}_{\nu\mu}. \tag{3.2}$$

Making use of the field transformation $\Phi = e^{-\phi}$, (3.1) can be rewritten as

$$S_G = \int d^4x\sqrt{-g}e^{-\phi}(R + \omega\,\phi^{,\alpha}\phi_{,\alpha}). \tag{3.3}$$

In what follows, we will regard the latter expression.

3.2.1 Integrable Weyl Geometry Through a Dynamical Way

Consider, in first place, the variation of (3.3) with respect to the connection. Assuming that the matter action S_m does not depend on it and that the independent connection is symmetric, the following expression is obtained

$$\nabla_\alpha(\sqrt{-g}e^{-\phi}g^{\mu\nu}) = 0. \tag{3.4}$$

By expanding the covariant derivative in (3.4) it is easy to see how the Palatini formalism applying to the Brans-Dicke action leads to the Weylian compatibility condition between the metric and the independent connection:

$$\nabla_\alpha g_{\mu\nu} = g_{\mu\nu}\phi_{,\alpha}, \tag{3.5}$$

where the scalar field ϕ is now interpreted as a geometrical field, namely, the Weyl scalar field.

In view of the fact that integrable Weyl geometry appears as the natural mathematical language to describe space-time, it is reasonable to consider that action (3.3) will vary as well, with respect to the scalar field ϕ. Therefore, we will propose an extension of Palatini variational method to derive the field equations, in the light of the fact that three independent geometrical entities should be considered here, namely, a Lorentzian metric $g_{\mu\nu}$ that measures length and angles, a symmetric connection Γ that is responsable by parallel transport and covariant derivatives of fields, and a scalar field ϕ that contributes to parallel transport of vectors, modifying their length in each point, defining a scale transformation.

Before we go further, some comments about Weylian geometry are in order. Broadly speaking, we can say that the geometry conceived by Weyl is a simple generalization of Riemannian geometry. Indeed, instead of regarding $\nabla_\alpha g_{\mu\nu} = 0$, we assume the more general compatibility condition $\nabla_\alpha g_{\mu\nu} = \sigma_\alpha g_{\mu\nu}$, where σ_α denotes the components of a one-form field. If $\sigma = d\phi$, where ϕ is a scalar field, then we have what is called an *integrable Weyl geometry*. In this sense, Eq. (3.5) represents the compatibility condition for this particular case. Analogously to Riemannian geometry, condition (3.5) is sufficient to determine the Weyl connection ∇ in terms of the metric g and the Weyl scalar field. Moreover, we can express the components of the affine connection completely in terms of the components of g and ϕ:

$$\Gamma^{\alpha}_{\mu\nu} = \left\{{}^{\alpha}_{\mu\nu}\right\} + \frac{1}{2}g^{\alpha\beta}\left(g_{\beta\mu}\phi_{,\nu} + g_{\beta\nu}\phi_{,\mu} - g_{\mu\nu}\phi_{,\beta}\right), \tag{3.6}$$

where $\left\{{}^{\alpha}_{\mu\nu}\right\}$ represents the Chistoffel symbols.

Let us give some definitions at this point. The set (M, g, ϕ) consisting of a differentiable manifold M endowed with a metric g and a Weyl scalar field ϕ. It is referred to as a *Weyl frame* [6]. It is interesting to note that condition (3.5) remains unchanged when we go to another Weyl frame $(M, \bar{g}, \bar{\phi})$ by performing the following simultaneous transformations

$$\bar{g}_{\mu\nu} = e^{-f}g_{\mu\nu}, \tag{3.7}$$

$$\bar{\phi} = \phi - f.$$

As a particular case, we have the *Riemann frame* which has a special interest. In such frame, the Weylian scalar field vanishes, and Riemann geometry is recovered. It is worth to mention that it is always possible to go to the Riemann frame, in which $\bar{\phi} = 0$. Indeed, we need only choose $f = \phi$ in the transformation Eq. (3.7).

As a result, we have the invariance of the affine connection coefficients $\Gamma^{\alpha}_{\mu\nu}$ under Weyl transformations, which in turn, implies the invariance of the affine geodesics.

3.2.2 How to Couple Weyl Geometry and Matter?

Before we consider the variation with respect to the metric, we need to specify the form of the matter action S_m. Let us show first why the matter action of the metric Brans-Dicke (BD) theory does not constitute a good choice. In view of the fact already established that space-time is in addition characterized by a scalar field, it is reasonable to expect that matter will couple to the metric as well as to the scalar field.

Another reason for rejecting the BD coupling for matter is the following: the matter action of the Brans-Dicke theory in the Palatini formalism does not satisfy the *Einstein Equivalence Principle* (EEP). This will be shown in detail in the next section.

Thus, our aim is to find out a matter action these actually satisfies that two requirements. For this purpose, let us rewrite condition (3.5) in a different form,

$$\nabla_{\alpha}(e^{\phi} g^{\mu\nu}) = 0. \tag{3.8}$$

Now it is useful to define the following 'effective' metric

$$\gamma^{\mu\nu} = e^{\phi} g^{\mu\nu}. \tag{3.9}$$

It is easy to see that the simplest matter action that involves coupling to the metric and the scalar field implies a minimal coupling to the effective metric (3.9):

$$\begin{aligned} S_m = \kappa \int d^4x \sqrt{-\gamma} L_m(\gamma_{\mu\nu}, \Psi, \nabla^{(\gamma)}\Psi) \\ = \kappa \int d^4x \sqrt{-g} e^{-2\phi} L_m(e^{-\phi} g_{\mu\nu}, \Psi, \nabla^{(e^{-\phi}g)}\Psi), \end{aligned} \tag{3.10}$$

where Ψ generically denotes the matter fields. In addition, as we will see, S_m defined as (3.10) assures that the EEP is satisfied.

We are now ready to perform the variation of (3.3) with respect to the metric $g_{\mu\nu}$, when (3.10) is taking into account. We then obtain the following field equations:

$$R_{\mu\nu} - \frac{1}{2} g_{\mu\nu} R = -\kappa T_{\mu\nu} - \omega \left(\phi_{,\mu}\phi_{,\nu} - \frac{1}{2} g_{\mu\nu}\phi^{,\alpha}\phi_{,\alpha} \right), \tag{3.11}$$

where it was assumed that the expression for the energy-momentum tensor is given by $T_{\mu\nu} = \frac{1}{\sqrt{-\gamma}}\frac{\delta(\sqrt{-\gamma}L_m)}{\delta\gamma^{\mu\nu}}$, from $\delta_g S = 0$.

Taking the trace of the Eq. (3.11) with respect to the metric $g_{\mu\nu}$ we obtain

$$R + \omega\phi^{,\alpha}\phi_{,\alpha} = \kappa T. \tag{3.12}$$

Let us now perform the variation of the action with respect to the scalar field ϕ. This leads to

$$R + 3\omega\phi^{,\alpha}\phi_{,\alpha} + 2\omega\Box\phi = \kappa T, \tag{3.13}$$

where $T = g^{\mu\nu}T_{\mu\nu}$ and we have used that

$$\delta_\phi S_m = \int d^4x\sqrt{-g}\frac{\delta(\sqrt{-\gamma}L_m)}{\delta\gamma^{\alpha\beta}}\frac{\delta\gamma^{\alpha\beta}}{\delta\phi}\delta\phi = -\int d^4x\sqrt{-g}e^{-\phi}T\delta\phi. \tag{3.14}$$

Finally, combining Eqs. (3.13) and (3.12) we obtain the field equation for the scalar field ϕ

$$\Box\phi + \phi^{,\alpha}\phi_{,\alpha} = 0, \tag{3.15}$$

where $\Box\phi = \phi^{;\alpha}_{;\alpha}$ is the d'Alembertian operator calculated with the Weylian connection. Rewriting the above equation in Riemannian terms and returning to the variable $\Phi = e^{-\phi}$ we have

$$\bar{\Box}\Phi = 0. \tag{3.16}$$

Note that Eq. (3.16) is analogous to the field equation for Φ in the *metric* BD theory for vacuum.

3.3 Conservation Law for Matter

Let us consider Eq. (3.11) and take its divergence. We thus will have

$$G^{\nu}_{\mu;\nu} = -\kappa T^{\nu}_{\mu;\nu} - \omega\left(\phi^{,\nu}\phi_{,\mu} - \frac{1}{2}\delta^{\nu}_{\mu}\phi^{,\alpha}\phi_{,\alpha}\right)_{;\nu}. \tag{3.17}$$

By using (3.15), (3.5) and Bianchi's identity, $G^{\nu}_{\mu;\nu} = -\phi_{,\nu}G^{\nu}{}_{\mu}$, in (3.17), we will obtain

$$\kappa T^{\nu}_{\mu;\nu} = \phi_{,\nu}G^{\nu}_{\mu} + \frac{\omega}{2}\phi_{,\mu}\phi_{,\beta}\phi^{,\beta}. \tag{3.18}$$

Making use of (3.11), we will finally get

$$T^{\nu}_{\mu;\nu} = -T^{\nu}{}_{\mu}\phi_{,\nu}, \tag{3.19}$$

or, in a succinct way,

$$(e^{\phi} T_{\mu}^{\ \nu})_{;\nu} = 0. \tag{3.20}$$

Now, we are ready to show that, as it has been mentioned above, the matter coupling (3.10) assures the EEP. Let us consider Eq. (3.20) for the energy-momentum tensor of dust[1], that is, $T_{\mu\nu} = \rho\gamma_{\mu\alpha}\gamma_{\nu\beta}u^{\alpha}u^{\beta}$, ρ being the energy density and $u^{\nu} = \frac{dx^{\nu}}{d\sigma}$ satisfies the parametrization condition $\gamma^{\mu\nu}u_{\mu}u_{\nu} = -1$. A straightforward calculation leads to the following equations

$$\frac{du^{\alpha}}{d\sigma} + \left(\left\{{}^{\alpha}_{\mu\nu}\right\} + \frac{1}{2}g^{\alpha\beta}\left(g_{\beta\mu}\phi_{,\nu} + g_{\beta\nu}\phi_{,\mu} - g_{\mu\nu}\phi_{,\beta}\right)\right)u^{\mu}u^{\nu} = 0, \tag{3.21}$$

which we readily identify as the Weylian affine geodesic equations.

On the other hand, the Brans-Dicke matter action, $S_m = \int d^4 \times \sqrt{-g}L_m$, leads to the following covariant divergence for the energy momentum tensor

$$T_{\mu;\nu}^{\nu\ (g)} = -\frac{\phi_{,\mu}}{2}T^{(g)} - 2\phi_{,\nu}T_{\mu}^{\nu(g)}, \tag{3.22}$$

with $T_{\mu\nu}^{(g)} = \frac{1}{\sqrt{-g}}\frac{\delta(\sqrt{-g}L_m)}{\delta g^{\mu\nu}}$. Considering again an energy-momentum tensor for dust, choosing $\gamma^{\mu\nu}u_{\mu}u_{\nu} = -1$ and taking into account that here we have $T^{(g)} = g^{\mu\nu}T_{\mu\nu}^{(g)} = -e^{-\phi}\rho$, it is not difficult to verify that the worldlines of the particles now must satisfy the equations

$$u^{\nu}u_{\mu;\nu} = \frac{e^{-\phi}}{2}(\dot{\phi}u_{\mu} + \phi_{,\mu}), \tag{3.23}$$

which clearly do not represent geodesic equations, hence implying that the EEP is not satisfied.

3.4 The Brans-Dicke Action in the Context of Weyl's Frame Formalism

Until now, we have developed our theory in the Weyl frame. To gain some insight, let us apply the Weyl transformations (3.7) to the action (3.3) with the purpose to

[1] It is a well known fact that the geodesic equations may be derived directly from the field equations. This is a known result, which goes back to Einstein and Papapetrou (see, for instance, [7–9]. The argument goes like this: Consider an assembly of free particles (i.e. not interacting with each other). If there are many of them, we can consider them together as a pressureless perfect fluid (dust). Then, a straightforward calculation shows that Eq. (3.20) leads to the affine geodesic equations with the connection coefficients given by Eq. (3.6) (of course, in the case of general relativity the scalar field ϕ does not appear in the equations). Therefore, since the world lines of the fluid particles are geodesics and because they are not interacting with each other we can infer that the world line of a single free particle is a geodesic.

achieve a Riemann frame, namely a frame in which we have $\bar{\phi} = 0$. Therefore, assuming $f = \phi$ and taking into account the invariance under (3.7) of the Riemann and the Ricci tensors and so $\bar{R} = \bar{g}^{\mu\nu}\bar{R}_{\mu\nu} = e^{-f}g^{\mu\nu}R_{\mu\nu} = e^{-f}R$, (3.3) results

$$S = \int d^4x\sqrt{-\gamma}(\bar{R}(\gamma, 0) + \omega\phi^{,\alpha}\phi_{,\alpha}) + S_m(\gamma, \Psi, \nabla^\gamma\Psi). \tag{3.24}$$

Due to the invariance of the matter action, the last term will preserve its form under this kind of transformations, that is, $S_m(\gamma, 0) = S_m(g, \phi)$. At this point, one could ask why a term involving ϕ still remains in the action in a frame where there is no Weyl field. As a matter of fact, after the Weyl transformation we have carried out, the remaining ϕ no longer represents the Weyl field, which completely vanishes in the new frame. The presence, in the action, of the term involving ϕ must be regarded as a mere trace left out by the specific Weyl transformation that we have considered, which implicit involves the scalar field. Thus, ϕ no longer plays a geometrical role, and accordingly must be interpreted as a physical field, in much the same way, as in the Brans-Dicke theory.

In view of (3.24) and from a redefinition of the scalar field, namely $\varphi = \sqrt{2\omega}\phi$, we conclude that our theory in the Weyl frame mapped by Weyl transformation into the Riemann frame leads to Einstein's theory of gravity minimally coupled to a massless scalar field.

Likewise, the fields equations in the Riemann frame are

$$\bar{R}_{\mu\nu} - \frac{1}{2}\gamma_{\mu\nu}\bar{R} = -\kappa T_{\mu\nu} - 2\left(\varphi_{,\mu}\varphi_{,\nu} - \frac{1}{2}\gamma_{\mu\nu}\varphi^{,\alpha}\varphi_{,\alpha}\right), \tag{3.25}$$

and

$$\bar{\Box}\varphi = 0, \tag{3.26}$$

where $\bar{\Box}$ is the d'Alembertian operator calculated using the Christoffel symbols with respect to $\gamma_{\mu\nu}$.

3.4.1 Experimental Tests on the Solar System

Here, we will briefly examine the classical tests in the solar system for the Brans-Dicke type theory when the description of the space-time is given by a Weyl integrable manifold. It is clear that the field equations in the Riemann frame appear in a simpler way than in the Weyl frame. Therefore, and taking advantage of the invariance of the geodesic equations under Weyl transformations, it is convenient to perform the tests on solar system in the Riemannian frame. Thus, let us consider the field Eqs. (3.25) and (3.26). As was already noted, performing a simple redefinition of the scalar field results easy to get exactly the same field equations as in the case of Einstein's theory of gravitation with massless scalar field, whose static, spherically symmetric, and asymptotically flat solution is given by [10, 11]

$$ds^2 = W(r)^S dt^2 - W(r)^{-S} dr^2 - r^2 W(r)^{1-S} d\Omega, \tag{3.27}$$

$$\phi = \frac{1}{\sqrt{2\omega}} \frac{\Sigma}{\eta} \ln |W(r)|, \tag{3.28}$$

$$W(r) = 1 - \frac{r_0}{r}, \tag{3.29}$$

where $S = \frac{M}{\eta}$, $r_0 = 2\eta$, $\eta = \sqrt{M^2 + \Sigma^2}$ and $M > 0$ is the body's mass in the center of this coordinates and Σ, its scalar charge. Note that we recover Schwarzchild's solution in the case $\Sigma = 0$. Using the parametrized post-Newtonian formalism it can be shown that in the first order approximation the Fisher space-time predicts the same effects in solar-system experiments as given by the Schwarzchild solution [12]. The invariance of the geodesic motion then guarantees that the same is true for the present scalar-tensor theory of gravitation in any Weyl frame.

3.5 Conclusion

Starting with the Brans-Dicke gravitational action and using the Palatini variational method to derive the field equations, we have presented a new scalar-tensor theory that reformulates the original BD theory in a non-Riemannian geometry, namely, the Weyl integrable geometry. In order to maintain valid the EEP, we had to modify the usual coupling between matter and geometry of General Relativity and Brans-Dicke theories. Since in the proposed formalism the scalar field is an inherent element of the geometry, we have assumed that the matter term must involve it in such a manner that assures that particles and light rays follow geodesics in the Weyl integrable space-time.

As a second result of the introduction of a new matter coupling, the corresponding conservation law for the energy-momentum tensor explicitly shows an exchange of energy between the Weyl field and matter fields. It has been noted that matter does not appear as source of the scalar field but both interact through the metric field. Another feature of the Weyl geometry is the invariance of geodesics under Weyl transformations, which implies that the classical tests can be made in the Riemann frame and are automatically satisfied in all other frames connected by a Weyl transformation.

To conclude, we mention that we are investigating the cosmological models in this theory, where we expect a different set of solutions from those in BD theory, since we have introduced a different coupling between matter and gravitational fields.

References

1. C. Will, *Theory and experiment in gravitational physics* (Cambridge University Press, Cambridge, 1993)
2. V. Faraoni, *Cosmology in Scalar-Tensor Gravity* (Kluwer Academic Publishers, Dordrecht, 2004)
3. M. Ferraris, M. Francaviglia, C. Reina, Gen . Rel. Grav. **14**, 243 (1982)
4. R.M. Wald, *General Relativity* (Chicago University Press, Chicago, 1984)
5. C. Brans, R. H. Dicke, Phys. Rev. **124**, 925 (1961)
6. C. Romero, J. B. Fonseca-Neto, M.L. Pucheu, Class. Quantum Grav. **29**, 155015 (2012)
7. A. Einstein, L. Infeld, B. Hoffmann, Ann. Math. Princeton, **39** (1938)
8. A. Papapetrou, Proc. Roy. Soc. **A209**, 248 (1951)
9. A. Papapetrou, E. Corinaldesi, Proc. Roy. Soc. **A209**, 259 (1951)
10. I. Z. Fisher, Zhurnal Experimental'noj i Teoreticheskoj Fiziki **18**, 636 (1948)
11. S. Abdolrahimi, A.A. Shoom, Phys. Rev. D **81**, 024035 (2010)
12. J. B. Formiga, Phys. Rev. D **83**, 087502 (2011)

Chapter 4
Gravitational Anomalies Signaling the Breakdown of Classical Gravity

X. Hernandez, A. Jiménez and C. Allen

Abstract Recent observations for three types of astrophysical systems severely challenge the GR plus dark matter scenario, showing a phenomenology which is what modified gravity theories predict. Stellar kinematics in the outskirts of globular clusters show the appearance of MOND type dynamics on crossing the a_0 threshold. Analysis shows a "Tully-Fisher" relation in these systems, a scaling of dispersion velocities with the fourth root of their masses. Secondly, an anomaly has been found at the unexpected scales of wide binaries in the solar neighbourhood. Binary orbital velocities cease to fall along Keplerian expectations, and settle at a constant value, exactly on crossing the a_0 threshold. Finally, the inferred infall velocity of the bullet cluster is inconsistent with the standard cosmological scenario, where much smaller limit encounter velocities appear. This stems from the escape velocity limit present in standard gravity; the "bullet" should not hit the "target" at more than the escape velocity of the joint system, as it very clearly did. These results are consistent with extended gravity, but would require rather contrived explanations under GR, each. Thus, observations now put us in a situation where modifications to gravity at low acceleration scales cease to be a matter of choice, to now become inevitable.

4.1 Introduction

Over the past few years the dominant explanation for the large mass to light ratios inferred for galactic and meta-galactic systems, that these are embedded within massive dark matter halos, has begun to be challenged. Direct detection of the dark matter particles, in spite of decades of extensive and dedicated searches, remains

X. Hernandez (✉)
Instituto de Astronomía, Universidad Nacional Autónoma de México,
AP 70-264, Distrito Federal, 04510, México
e-mail: xavier@astroscu.unam.mx

A. Jiménez · C. Allen
Instituto de Astronomía, Universidad Nacional Autónoma de México,
Distrito Federal, México
e-mail: mjimenez@astro.unam.mx

C. Allen
e-mail: chris@astroscu.unam.mx

C. Moreno González et al. (eds.), *Accelerated Cosmic Expansion*,
Astrophysics and Space Science Proceedings 38,
DOI 10.1007/978-3-319-02063-1_4,

lacking. This has led some to interpret the velocity dispersion measurements of stars in the local dSph galaxies (e.g. [1, 17]), the extended and flat rotation curves of spiral galaxies (e.g. [41]), the large velocity dispersions of galaxies in clusters (e.g. [39]), stellar dynamics in elliptical galaxies (e.g. [40]), the gravitational lensing due to massive galaxies (e.g. [10, 34, 53]), and even the cosmologically inferred matter content for the universe through CMB and structure formation physics (e.g. [16, 39, 48]), not as indirect evidence for the existence of a dominant dark matter component, but as direct evidence for the failure of the current Newtonian and General Relativistic theories of gravity, in the large scale or low acceleration regimes relevant for the above situations.

Numerous alternative theories of gravity have recently appeared (e.g. [5], and variations; [7, 49], F(R) theories e.g. [8, 9], conformal gravity theories e.g. [29]), mostly grounded on geometrical extensions to General Relativity, and leading to laws of gravity which in the large scale or low acceleration regime, mimic the MOdified Newtonian Dynamics (MOND) fitting formulas. Similarly, [33] have explored MOND not as a modification to Newton's second law, but as a modified gravitational force law in the Newtonian regime, finding a good agreement with observed dynamics across galactic scales without requiring dark matter. In fact, recently [4] have constructed an $f(R)$ extension to general relativity which in the low velocity limit converges to the above approach.

A generic feature of all of the modified gravity schemes mentioned above is the appearance of an acceleration scale, a_0, above which classical gravity is recovered, and below which the dark matter mimicking regime appears. This last feature results in a general prediction; all systems where $a >> a_0$ should appear as devoid of dark matter, and all systems where $a << a_0$ should appear as dark matter dominated, when interpreted under classical gravity. It is interesting that no $a >> a_0$ system has ever been detected where dark matter needs to be invoked, in accordance with the former condition. On the other hand, the latter condition furnishes testable predictions. First, notice that for test particles in orbit around a $1M_{\odot}$ star, in circular orbits of radius s, the acceleration is expected to drop below $a_0 \approx 1.2 \times 10^{-10}\, m/s^2$ for $s >$7000 AU= $3.4 \times 10^{-2} pc$. The above provides a test for the dark matter/modified theories of gravity debate; the relative velocities of components of binary stars with large physical separations should deviate from Kepler's third law under the latter interpretation.

More specifically, seen as an equivalent Newtonian force law, beyond $s \approx 7000$ AU the gravitational force should gradually switch from the classical form of $F_N = GM/s^2$ to $F_{MG} = (GMa_0)^{1/2}/s$, and hence the orbital velocity, $V^2/s = F$, should no longer decrease with separation, but settle at a constant value, dependent only on the total mass of the system through $V = (GMa_0)^{1/4}$. That is, under modified gravity theories, binary stars with physical separations beyond around 7000 AU should exhibit "flat rotation curves" and a "Tully-Fisher relation", as galactic systems in the same acceleration regime do.

In [18] we proposed that wide binary orbits may be used to test Newtonian gravity in the low acceleration regime. There we applied this test to the binaries of two very recent catalogues containing relative velocities and separations of wide binaries. The two catalogues are entirely independent in their approaches.

At the somewhat larger scales of globular clusters, with sizes of tens of parsecs and masses of order $10^5 M_{\odot}$, the central values of the stellar velocity dispersion, projected on the plane of the sky, for many Galactic globular clusters (GC) have been well known for decades, and are known to accurately correspond to the expectations of self-consistent dynamical models under Newtonian gravity, e.g. King models (e.g. [6]). Recently, a number of studies (e.g. [23–26, 43–46], henceforth the Scarpa et al. and Lane et al. groups respectively) have performed measurements of the projected velocity dispersion along the line of sight for stars in a number of Galactic GCs, but as a function of radius, and reaching in many cases out to radial distances larger than the half-light radii of the clusters by factors of a few.

The surprising result of the above studies has been that radially, although velocity dispersion profiles first drop along Newtonian expectations, after a certain radius, settle to a constant value, which varies from cluster to cluster. As already noted by [46], it is suggestive of a modified gravity scenario that the point where the velocity dispersion profiles flatten, approximately corresponds to the point where average stellar accelerations drop below a_0. Several recent studies have shown dynamical models for self-gravitating populations of stars under MOND or other modified gravity variants (e.g. [14, 15, 20, 36, 50]) which accurately reproduce not only the observed velocity dispersion profiles, but also the observed surface brightness profiles.

From the point of view of assuming Newtonian gravity to be exactly valid at all low velocity regimes, it has also been shown that both velocity dispersion and surface brightness profiles for Galactic GCs can be self-consistently modelled. Under this hypothesis, it is dynamical heating due to the overall Milky Way potential that is responsible for the flattening of the velocity dispersion profiles e.g. [12, 24]. The constant velocity dispersion observed at large radii merely shows the contribution of unbound stars in the process of evaporating into the Milky Way halo. In attempting to sort between these two contrasting scenarios, we took a fully empirical approach in [19]. There, we critically examine the plausibility of both gravitational scenarios by looking through the data for other correlations which each suggest.

For the Newtonian case, we examine the best available inferences for the tidal radius of each cluster at closest galacto-centric passage, and compare it to the observed point where the velocity dispersion flattens. We found the former to generally exceed the latter by factors of 4 on average, making the Newtonian interpretation suspect. Also, we take all the clusters which the Lane et al. group have claimed show no indication of a modified gravity phenomenology, based on the fact that their velocity dispersion profiles can be modelled using Plummer profiles, and show that the fits with the generic asymptotically flat profiles we use are actually slightly better, in all cases.

We shall use the term MONDian to refer to any modified theory of gravity which reproduces the basic phenomenology of MOND in the low velocity limit for accelerations below a_0, of flat equilibrium velocities and a Tully-Fisher relation, regardless of the details of the fundamental theory which might underlie this phenomenology.

Together with our previous results of [20] showing that the asymptotic values of the velocity dispersion profiles are consistent with scaling with the fourth root of the total masses, a Tully-Fisher relation for GCs, our results support the interpretation of the observed phenomenology as evidence for a change in regime for gravity on crossing the a_0 threshold.

A recent study reaching the same conclusions, but at a significantly distinct scale, can be found in, [28] show that the infall velocity of the two components of the Bullet cluster, as required to account for the hydrodynamical shock observed in the gas, is inconsistent with expectations of full cosmological simulations under standard ΛCDM assumptions. This has recently been confirmed at greater detail by [51], and can in fact be seen as a failure not only of the ΛCDM model, but of standard gravity, as the required collisional velocity is actually larger than the escape velocity of the combined system. We note also the recent reviews by [22] and references therein, detailing a number of observations in tension with standard ΛCDM assumptions.

4.2 Dynamics of Wide Binaries

The Newtonian prediction for the relative velocities of the two components of binaries having circular orbits, when plotted against the binary physical separation, s, is for a scaling of $\Delta V \propto s^{-1/2}$, essentially following Kepler's third law, provided the range of masses involved were narrow. In a relative proper motion sample however, only two components of the relative velocity appear, as velocity along the line of sight to the binary leads to no proper motion. Thus, orbital projection plays a part, with systems having orbital planes along the line of sight sometimes appearing as having no relative proper motions. A further effect comes from any degree of orbital ellipticity present; it is hence clear that the trend for $\Delta V \propto s^{-1/2}$ described above, will only provide an upper limit to the distribution of projected ΔV vs. s expected in any real observed sample, even if only a narrow range of masses is included. One should expect a range of measured values of projected ΔV at a fixed observed projected s, all extending below the Newtonian limit, which for equal mass binaries in circular orbits gives $\Delta V_N = 2\left(\frac{GM}{s}\right)^{1/2}$.

Further, over time, the orbital parameters of binaries will evolve due to the effects of Galactic tidal forces and dynamical encounters with other stars in the field, specially in the case of wide binaries. To first order, one would expect little evolution for binaries tighter than the tidal limit of 1.7 pc, and the eventual dissolution of wider systems. A very detailed study of all these points has recently appeared, [21]. These authors numerically follow populations of 50,000 $1M_\odot$ binaries in the Galactic environment, accounting for the evolution of the orbital parameters of each due to the cumulative effects of the Galactic tidal field at the Solar radius. Also, the effects of close and long range encounters with other stars in the field are carefully included, to yield a present day distribution of separations and relative velocities for an extensive population of wide binaries, under Newtonian Gravity.

It is found that when many wide binaries cross their Jacobi radius, the two components remain fairly close by in both coordinate and velocity space. Thus, in any real wide binary search a number of wide pairs with separations larger than their Jacobi radii will appear. Finally, [21] obtain the RMS one-dimensional relative velocity difference, ΔV_{1D}, projected along a line of sight, for the entire populations of binaries dynamically evolved over 10 Gyr to today, as plotted against the projected separation on the sky for each pair. The expected Keplerian fall of $\Delta V_{1D} \propto s^{-1/2}$ for separations below 1.7 pc is obtained, followed by a slight rise in ΔV_{1D} as wide systems cross the Jacobi radius threshold. ΔV_{1D} then settles at RMS values of $\approx 0.1\ km/s$. This represents the best currently available estimate of how relative velocities should scale with projected separations for binary stars (both bound and in the process of dissolving in the Galactic tides) under Newtonian gravity. We see that all we need is a large sample of relative proper motion and binary separation measurements to test the Newtonian prediction for the RMS values of the 1 dimensional relative velocities of [21], and to contrast the $\Delta V_N \propto s^{-1/2}$ and the $\Delta V_{MG} = cte.$ predictions for the upper envelope of the ΔV vs. s distributions.

In the [47] catalogue wide binaries are identified by assigning a probability above chance alignment to the systems by carefully comparing to the underlying background (and its variations) in a 5 dimensional parameter space of proper motions and spatial positions. We keep only binaries with a probability of non-chance alignment greater than 0.9. The binary search criteria used by the authors requires that the proposed binary should have no near neighbours; the projected separation between the two components is thus always many times smaller than the typical interstellar separation. We use the reported distances to the primaries, where errors are smallest, to calculate projected ΔV and projected s from the measured $\Delta\mu$ and $\Delta\theta$ values reported. Although the use of *Hipparcos* measurements guarantees the best available quality in the data, we have also removed all binaries for which the final signal to noise ratio in the relative velocities was lower than 0.3 (Figs. 4.1 and 4.2).

We are left with a sample of 280 binaries, having distances to the Sun within $6 < d < 100$ in pc. The data show a perfectly flat upper envelope in a ΔV vs. projected s, [20]. The average signal to noise ratio for the data is 1.7, with an average error on ΔV of 0.83 km/s, which considering a 2σ factor from the top of the distribution to the real underlying upper limit for the sample, results in 3 km/s as our estimate of the actual physical upper limit in ΔV.

The Sloan low mass wide pairs catalogue (SLoWPoKES) of [11] contains a little over 1,200 wide binaries with relative proper motions for each pair, distances and angular separations. Also, extreme care was taken to include only physical binaries, with a full galactic population model used to exclude chance alignment stars using galactic coordinates and galactic velocities, resulting in an estimate of fewer than 2 % of false positives. This yields only isolated binaries with no neighbours within many times the internal binary separation. Again, we use the reported distances to the primaries to calculate projected ΔV and projected s from the measured $\Delta\mu$, $\Delta\theta$ and d values reported by [11], to obtain a sample of 417 binaries.

The upper envelope of the distribution of ΔV from this catalogue does not comply with Kepler's third law. As was the case with the *Hipparcos* sample, the upper

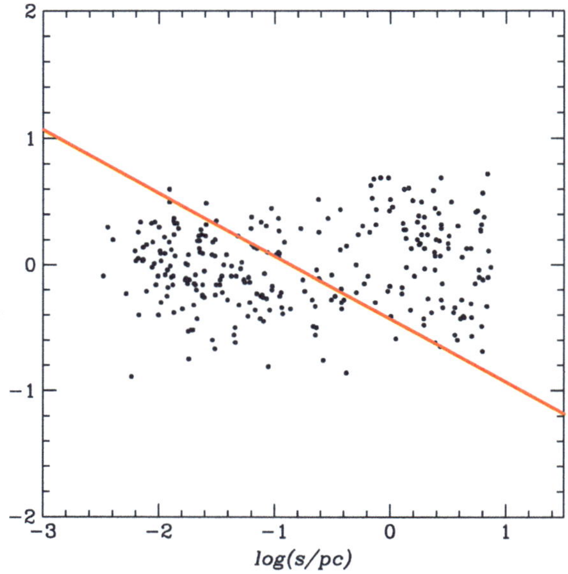

Fig. 4.1 The figure shows projected relative velocities and separations for each pair of wide binaries from the [47] *Hipparcos* catalogue having a probability of being the result of chance alignment < 0.1. The average value for the signal to noise ratio for the sample shown is 1.7. The upper limit shows the flat trend expected from modified gravity theories, at odds with Kepler's third law, shown by the $s^{-1/2}$ *solid line*

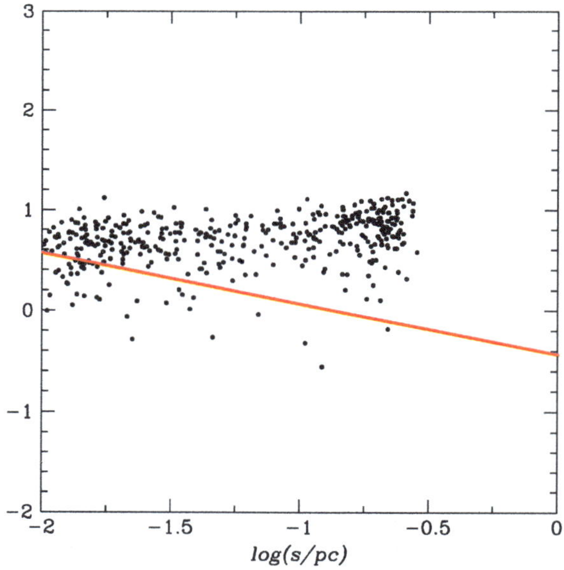

Fig. 4.2 The figure shows projected relative velocities and separations for each pair of wide binaries from the [11] SDSS catalogue within the distance range $(225 < d/pc < 338)$. The average value for the signal to noise ratio for the sample shown is 0.5. The upper limit shows the flat trend expected from modified gravity theories, at odds with Kepler's third law, shown by the $s^{-1/2}$ *solid line*

envelope describes a flat line, as expected under modified gravity schemes. The average signal to noise in ΔV for the [11] catalogue is 0.48, with an average error on ΔV of 12 km/s, which considering a 2σ factor from the top of the complete distribution to the real underlying upper limit gives the same 3 km/s as obtained for the [47] *Hipparcos* catalogue.

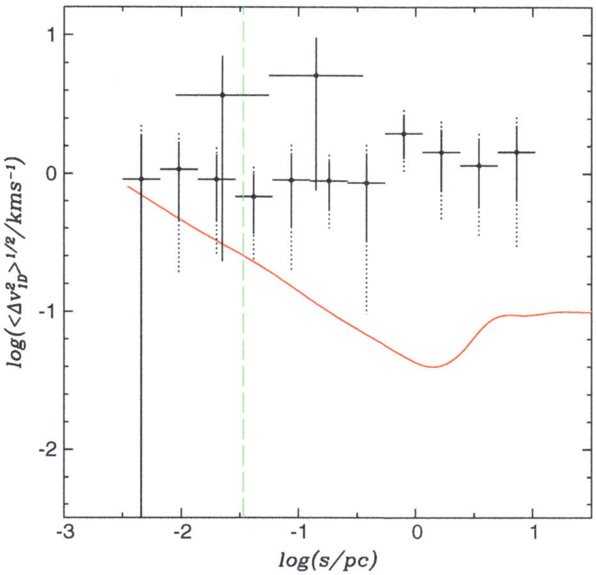

Fig. 4.3 The solid curve gives the RMS values for one dimensional projected relative velocities as a function of projected separations, for the detailed dynamical modelling of large populations of wide binaries evolving in the Galactic environment, taken from [21]. The same quantity for the data from the catalogues analysed is given by the points with error bars; those with narrow *log(s)* intervals being from the *Hipparcos* sample of [47], and those two with wide *log(s)* intervals coming from the SDSS sample of [11]

Figure 4.3 shows the RMS value of the one-dimensional relative velocity difference for both of the samples discussed. The error bars give the error propagation on $\Delta\mu$ and d. We construct ΔV_{1D} by considering only one coordinate of the two available from the relative motion on the plane of the sky. Thus, each binary can furnish two ΔV_{1D} measurements, which statistically should not introduce any bias. Indeed, using only $\Delta\mu_l$ or only $\Delta\mu_b$ or both for each binary, yields the same mean values for the points shown. The small solid error bars result from considering an enlarged sample where each binary contributes two ΔV_{1D} measurements, while the larger dotted ones come from considering each binary only once, and do not change if we consider only $\Delta\mu_l$ or only $\Delta\mu_b$. The series of small $log(s)$ interval data are for the *Hipparcos* catalogue of [47], while the two broader crosses show results for the [11] SDSS sample.

The solid curve is the Newtonian prediction of the full Galactic evolutionary model of [21] for binaries, both bound and in the process of dissolving. Note that the results of this simulation deviate from Kepler's law for s larger than the Newtonian Jacobi radius of 1.7 pc, whereas the discrepancy with the observed samples also occurs at much smaller separations. Even considering the large error bars, where each binary contributes only one ΔV_{1D} value, we see eight points lying beyond 1σ, making the probability of consistency between this prediction and the observations of less than $(0.272)^8 = 3 \times 10^{-5}$.

We obtain a constant RMS value for ΔV_{1D} of 1 km/s, in qualitative agreement with expectations from modified gravity schemes. The vertical line marks $a = a_0$; we see the data departing from the Newtonian prediction outwards of this line, and not before. The two independent catalogues, each using different sets of selection criteria, each perhaps subject to its own independent systematics, are consistent with the same result, a constant horizontal upper envelope for the distribution of relative velocities on the plane of the sky at an intrinsic value of 3 km/s $\pm$ 1 km/s, extending over 3 orders of magnitude in s, with a constant RMS ΔV_{1D} value consistent with 1 km/s $\pm$ 0.5 km/s. This supports the interpretation of the effect detected as the generic prediction of modified gravity theories.

4.3 Outer Dynamics of Globular Clusters

We begin by modelling the observed projected radial velocity dispersion profiles, $\sigma_{obs}(R)$, for the globular clusters in our sample, [19]. As seen from the Scarpa et al. and Lane et al. data, the observed velocity dispersion profiles show a central core region where the velocity dispersion drops only slightly, followed by a "Keplerian" zone where the drop is more pronounced. These first two regions are in accordance with standard Newtonian King profiles, but they are then followed by a third outermost region where the velocity dispersion profiles cease to fall along Keplerian expectations, and settle to fixed values out to the last measured point. As some of us showed in [20], an accurate empirical modelling for these velocity dispersion profiles can be achieved through the function:

$$\sigma(R) = \sigma_1 e^{-(R/R_\sigma)^2} + \sigma_\infty \tag{4.1}$$

In the above equation σ_∞ is the asymptotic value of $\sigma(R)$ at large radii, R_σ a scale radius fixing how fast the asymptotic value is approached, and σ_1 a normalisation constant giving $\sigma(R=0) = \sigma_1 + \sigma_\infty$.

We now take the observed data points $\sigma_{obs}(R_i)$ along with the errors associated to each data point, to determine objectively through a maximum likelihood method the best fit values for each of the three parameters in Eq. (4.1), for each of the 16 observed globular clusters. The confidence intervals for each of the three parameters are then obtained without imposing any marginalisation. This last point allows to properly account for any correlations between the three fitted parameters when calculating any quantity derived from combinations of them, as will be constructed in what follows.

Taking $\sigma_{obs}(R_i)$ data from [12, 23–27, 42–46], and half-light radii, $R_{1/2}$, from integrating the surface density brightness profiles of [52], we perform a maximum likelihood fit for all the sixteen globular clusters studied.

Figure 4.4 shows the observed projected velocity dispersion profiles for 6 representative globular clusters from our sample, points with error bars. The radial coordinate has been normalised to the $R_{1/2}$ radius of each of the clusters. The continuous curves show the maximum likelihood fits for each cluster, which are clearly good representations of the data. We can now give $R_f = 1.5 R_\sigma$ as an adequate empirical estimate of the radius beyond which the dispersion velocity profile becomes essentially flat. In terms of Eq. (4.1), which can be seen to be highly consistent with the observed velocity dispersion profiles, R_f is the radius such that $\sigma(R_f) = 0.1\sigma_1 + \sigma_\infty$,

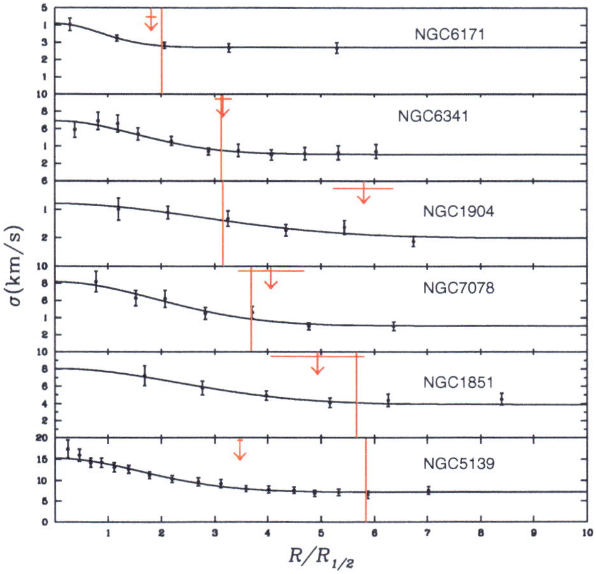

Fig. 4.4 The figure shows the observed projected velocity dispersion profiles for six representative GCs in our sample, points with error bars, as a function of the radial coordinate, normalised to the half-light radius of each. The solid curves give the maximum likelihood fits to the asymptotically flat $\sigma(R)$ model of Eq. (4.1), seen to be accurate descriptions of the data. The vertical lines indicate the $a = a_0$ threshold, and the arrows the point where the profiles flatten, *a priori* independent features, in most cases seen to occur at approximately the same place

a good representation of the transition to the flat behaviour, as can be checked from Fig. 4.4, where the arrows give R_f, with the horizontal lines on the arrows showing the 1σ confidence intervals on these fitted parameters. An empirical definition of the radius where the typical acceleration felt by stars drops below a_0 can now be given as R_a, where:

$$\frac{3\sigma(R_a)^2}{R_a} = a_0. \tag{4.2}$$

Using the above definition, we can now identify R_a for each of the globular clusters studied. The vertical lines in Fig. 4.5 show R_a for each cluster, also normalised to the half-light radius of each. In the figure, clusters have been ordered by their $R_a/R_{1/2}$ values, with the smallest appearing at the top, and $R_a/R_{1/2}$ growing towards the bottom of the figure.

The good fits shown in the studies of the Lane et al. group to the observed velocity dispersion profiles using Plummer models are clearly not sufficient to dismiss a modified gravity interpretation, as the asymptotically flat projected dispersion velocity fits of the type used for full dynamical modelling under modified gravity [20] actually provide even slightly better fits to the data, [19].

It is interesting at this point to notice a first correlation, the smaller the value of $R_a/R_{1/2}$, the larger the fraction of the cluster which lies in the $a < a_0$ regime, and interestingly, the flatter the velocity dispersion profile appears. At the top of the figure we see clusters where stars experience accelerations below a_0 almost at all radii, and it so happens, that it is only in these systems that the velocity dispersion profile appears almost flat throughout. Towards the bottom, we see systems where

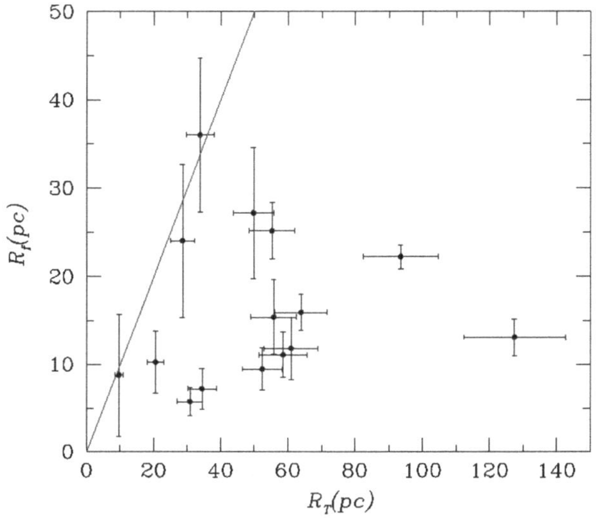

Fig. 4.5 The figure shows the relation between the point where the velocity dispersion flattens, R_f, and the Newtonian tidal radius, R_T, for each cluster. Even considering the large errors involved on both quantities, most points fall far to the right of the identity line shown, making the Newtonian explanation for the flattened velocity dispersion profiles, rather suspect

only at the outskirts accelerations fall under a_0. Over most of their extents, these clusters lie in the Newtonian $a > a_0$ regime, and indeed, it is exclusively these, that show a clear Keplerian decline in the projected velocity dispersion profiles over most of their extents. Also, notice that R_f and R_a approximately coincide, as already previously noticed by [43], the flattening in the velocity dispersion profiles seems to appear on crossing the a_0 threshold.

In order to test the validity of the explanation for the outer flattening of the observed velocity dispersion profiles under Newtonian gravity, that these indicate dynamical heating due to the tides of the Milky Way system (bulge plus disk plus dark halo), we need accurate estimates of the Newtonian tidal radii for the clusters studied. One of us in [2, 3] performed detailed orbital studies for 54 globular clusters for which absolute proper motions and line of sight velocities exist. In that study, both a full 3D axisymmetric Newtonian mass model for the Milky Way and a model incorporating a galactic bar were used to compute precise orbits for a large sample of globular clusters, which fortunately includes the 16 of our current study. The Galactic mass models used in those papers are fully consistent with all kinematic and structural restrictions available. Having a full mass model, together with orbits for each globular cluster, allows the calculation of the Newtonian tidal radius, not under any "effective mass" approximation, but directly through the calculation of the derivative of the total Galactic gravitational force, including also the evaluation of gradients in the acceleration across the extent of the clusters, at each point along the orbit of each studied cluster.

The Newtonian tidal radii we take for our clusters, R_T, are actually the values which results in the largest dynamical heating effect upon the clusters studied, those at perigalacticon. As the distance of closest approach to the centre of the Galaxy might vary from passage to passage, as indeed it often does, detailed orbital integration is used to take R_T as an average for perigalactic passages over the last 1 Gyr.

We update the tidal radii published in [2, 3], by considering revised total masses from the integration of the observed V band surface brightness profiles for our clusters [52], and using the V band stellar M/L values given in [32] and accompanying electronic tables. For each individual GC, detailed single stellar population models tuned to the inferred ages and metallicities of each of the clusters we model were constructed in that study, using various standard population synthesis codes, and for a variety of assumed IMFs. In this way present stellar M/L values in the V band were derived, which we used. As we did not in any way use the dynamical mass estimates of [32], the total masses we used are independent of any dynamical modelling or assumption regarding the law of gravity, as they are derived through completely independent surface brightness profile measurements and stellar population modelling. The confidence intervals in our tidal radii include the full range of stellar M/L values given by [32], through considering a range of ages, metallicities and initial mass functions consistent with the observed HR diagrams for each cluster.

In Fig. 4.5 we show values of R_f for our clusters, plotted against their corresponding R_T values, both in units of pc. The error bars in R_f come from full likelihood analysis in the fitting process of Eq. (4.1) to $\sigma_{obs}(R)$, which guarantees that confidence intervals in both of the quantities plotted are robust 1σ ranges. The solid line shows a $R_f = R_T$ relation. It is obvious from the figure that the onset of the flat velocity dispersion regime occurs at radii substantially smaller than the tidal radii, for all of the globular clusters in our sample. Even under the most extreme accounting of the resulting errors, only three of the clusters studied are consistent with $R_T \approx R_f$ at 1σ. Actually, the average values are closer to $R_T = 4R_f$, with values higher than 8 appearing. One of the clusters, NGC 5024 does not appear, as it has values of $R_T = 184.12$, $R_f = 36$, which put it out of the plotted range, but consistent with the description given above. Given the R^3 scaling of Newtonian tidal phenomena, even a small factor of less than 2 inwards of the tidal radii, tides can be safely ignored, e.g. in Roche lobe overflow dynamics, the stellar interior is largely unaffected by the tidal fields, until almost reaching the tidal radius. It therefore appears highly unlikely under a Newtonian scheme, that Galactic tides could be responsible for any appreciable dynamical heating of the velocity dispersion of the studied clusters.

We note that [24, 25, 27] find that Newtonian tidal heating can explain the observed velocity dispersion profile of their GC sample. However, it is important to note that in [24, 25, 27], total masses were calculated directly from the observed velocity dispersion observations, under the assumption that Newtonian dynamics hold. If that assumption is to be tested, the importance of deriving total masses through an independent method, not based on stellar dynamics, is evident.

Notice also, that most of the clusters in our sample are problematic for a Newtonian gravity scheme, even without the recent observations of an outer flat velocity dispersion profile. As remarked already in [2], the clusters in our sample have Newtonian tidal radii larger than the observed truncation radii of their light distribution, the sole exceptions being Omega Cen (NGC 5139) and M92 (NGC 6341), two rather anomalous clusters. Whereas a full dynamical modelling under an extended gravity force law of these clusters, [20], naturally yielded an outer truncation for the light profile, under a Newtonian hypothesis, the observed truncation in the light profile of the clusters in our sample cannot be explained as arising from interaction with the tidal field of the Milky Way.

Furthermore, notice that we have taken R_T at perigalacticon, where tides are at their most severe over the clusters orbit, any other orbital occupation averaging would result in substantially larger R_T values. Notice also that as shown by [2, 3], the inclusion of a realistic massive Galactic bar potential, in the case of the clusters in our sample, results generally in negligible changes in the resulting R_T values, or in some cases, a slight increase in these values. Hence, even taking the fullest non-axisymmetric Galactic mass model under Newtonian gravity, with precise orbits derived from 3D velocity measurements for the clusters studied, together with total mass determinations tuned to the individual stellar populations of them, yields tidal radii as shown in Fig. 4.5.

As already noticed by [46], the flattening in the observed velocity dispersion profiles seems to appear at the point where the a_0 threshold is crossed. Here we use the much more careful and objective modeling of the observed velocity dispersion curves of the previous section to test this point.

We see that most GCs in the sample fall within 1σ of the identity, [19], with about a third falling further away. A quantitative test of the correlation being explored is possible, since the careful modelling of the velocity dispersion profiles we performed naturally yields objective confidence intervals for the parameters of the fit.

We end this section with Fig. 4.6, which shows the relation between the measured asymptotic velocity dispersion, σ_∞, and the total mass of the clusters in question. These masses represent the best current estimates of the stellar mass for each of the clusters in the sample, including corresponding confidence intervals. As with all the other correlations and data presented, there is no dynamical modelling or modified gravity assumptions going into Fig. 4.6, merely observable quantities. We see, as already pointed out in [20], the GCs observed nicely comply with a scaling of $\sigma \propto M^{1/4}$, the Tully-Fisher law of galactic systems "embedded within massive dark haloes".

The straight line shows the best fit $\sigma \propto M^{1/4}$ scaling, and actually falls only a factor of 1.3 below the modified gravity prediction for systems lying fully within the low acceleration regime (e.g. [20], for the same value of $a_0 = 1.2 \times 10^{-10} m/s$ used here, as calibrated through the rotation curves of galactic systems. This small offset is not surprising, since the GCs treated here are not fully within the $a < a_0$ condition, most have an inner Newtonian region encompassing a substantial fraction of their masses. A preliminary version of this last figure appeared already in [20]; we reproduce here an updated version using now the extended sample of clusters treated, and σ_∞ values and their confidence intervals as derived through the careful velocity dispersion fitting procedure introduced.

To summarise, we have tested the Newtonian explanation of Galactic tides as responsible for the observed $\sigma(R)$ phenomenology, and found it to be in tension with the observations, given the tidal radii (at perigalacticon) which the GCs in our sample present, are generally larger than the points where $\sigma(R)$ flattens, on average,

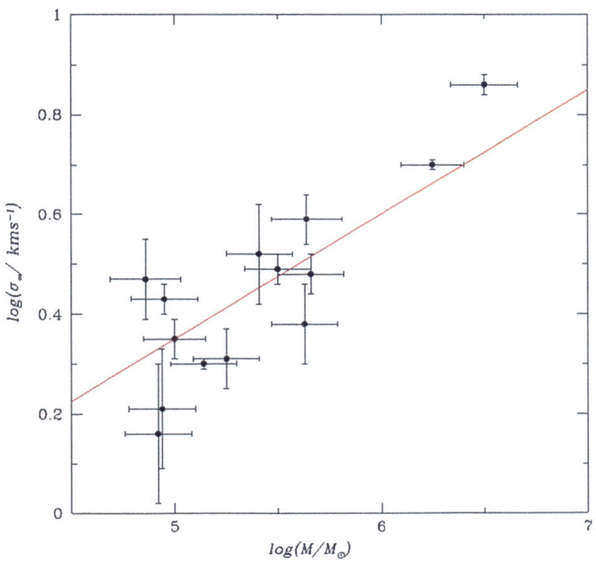

Fig. 4.6 Here we give the relation between the observed asymptotic dispersion velocity measurements, and the total mass of each cluster. The line gives the best fit $\sigma \propto M^{1/4}$ scaling for the data

by factors of 4, with values higher than 8 also appearing. An explanation under a MONDian gravity scheme appears likely, given the clear correlations we found for the clusters in our sample, all in the expected sense.

4.4 The Bullet Cluster and Other Inconcistencies of GR at Galactic Scales

In going to galactic scales, the local dwarf spheroidal satellites (dSph) of the Milky Way offer an interesting test bed for modified theories of gravity, being characterised by the highest dark matter fractions, when modelled under standard gravity. In [17] we showed that fully self consistent dynamical models can be constructed for these objects under a modified Newtonian force law, and found an interesting correlation. Assuming standard gravity, dSphs with the oldest stellar populations show the highest dark matter fractions, something which has to be regarded as a curious coincidence. However, under the assumption that the stars alone determine the gravitational potential of modified gravity schemes, it is natural to expect, indeed it is predicted, that the gravitational forces will be strongest, and deviate further from the Newtonian predictions, for the oldest stellar populations, having higher intrinsic mass to light ratios.

More recently, [22] has pointed out a serious inconsistency in the standard gravity interpretation of tidal dwarf galaxies. These small systems are formed during the interaction and merger processes of large galaxies. Within standard gravity, they are seen as transient stellar structures displaying out of equilibrium dynamics and containing no dark matter. This last condition follows from the large velocity dispersion values of the dark matter in the haloes of the large colliding galaxies, more than an order of magnitude larger than what would be required to form a stable halo around the forming small tidal dwarf galaxies. However, recent observations have confirmed that these tidal dwarf galaxies have the same dynamics as normal, stable dwarf galaxies showing relaxed, equilibrium dynamics "dominated by dark matter halos". Basically, tidal dwarf galaxies lie along the same Tully-Fisher relation as defined by normal galaxies, including more standard dwarfs. Thus, under a standard gravity plus dark matter interpretation, we must accept as a curious coincidence that tidal dwarf galaxies, transient swirls of tidally drawn material in interacting galaxies, have the same internal velocities as normal dwarf galaxies, equilibrium systems dominated by dark matter haloes. Under a MONDian modified view however, the observation above is a prediction; once tides form low density ensembles of stars, these will behave just like any other such set of stars, a standard dwarf galaxy dominated by the modified gravity regime, as internal accelerations are below a_0.

At the even largest Mpc scales of the bullet cluster, a serious inconsistency in the standard gravity interpretation has been pointed out by [28]. These authors show that the encounter velocities for the two components of the bullet cluster, as determined by the careful hydrodynamical modelling of [30] constrained to yield the observed bow shock in the glowing x-ray gas, are so high, that they are totally incompatible with the current standard cosmological scenario. This conclusion has recently been confirmed in much more detail by [51], who show clearly through full standard cosmological simulations, that the probability of finding such a system as the observed bullet cluster, at the observed redshift, is lower than 5×10^{-8}, essentially a statement of the impossibility of producing such a system under the standard cosmological scenario. The origin of these inconsistency is clear, and stems from the relation between the sound speed for the gas inside a virialized halo, and the escape velocity of these same halo. This last velocity is the highest at which an object falling into the central more massive cluster can be expected to appear, and it is only a factor of 2 higher than the isothermal sound speed in the gas. Thus, encounters with Mach numbers higher than two are inconsistent with standard gravity. Within an expanding cosmological scenario, one must first overcome the expansion, making the highest Mach numbers possible, somewhat lower than two. The strong bow shock seen in the x-ray gas of the bullet cluster however, testifies to an encounter with a Mach number higher than 2, and hence inconsistent with standard GR and Newtonian gravity. Clearly, no amount of dark matter can alleviate the problem, as adding dark matter will increase the impact velocity, but increase also, in the same proportion, the sound speed in the gas before the collision, and hence leave the Mach number of the shock unchanged. One can not simply say that both original clusters were orbiting within the higher gravitational potential of a much larger structure, as at that redshift of 0.5, in general, no larger structures excised, indeed, none is seen in the vicinity of the cluster in question.

4.5 Conclusions

We have reviewed recent dynamical observations for wide binaries in the solar neighbourhood which show clear departures from the Newtonian predictions on crossing the a_0 threshold, with equilibrium velocities becoming constant, as expected under MONDian modified gravity schemes calibrated to explain rotation curves of large galaxies without the need for any dark matter. Under a standard gravity interpretation, adding dark matter is not an option, as the scales involved are smaller than any where dark matter is expected, and any such addition would have to be extremely fine tuned; no dark matter for binaries having accelerations higher than a_0, and an increasing mini halo necessary as the semi-major axis grows beyond 7000 AU.

We have shown that the velocity dispersion profiles of globular clusters in the Milky Way tend to constant velocity values at the outskirts, at values which happen to scale precisely along the same $M^{1/4}$ relation as the galactic Tully-Fisher relation. This is natural under any MONDian gravity scheme, but is inconsistent with standard gravity, where tidal velocity disruptions show an inverse scaling with mass, and clear scalings with orbital parameters and internal concentrations, none of which appear in the globular clusters studied.

At the largest scales of the colliding galaxy clusters in the bullet cluster system, it has been shown that the standard cosmological scenario is incapable of producing the strong bow shock seen in the glowing x-ray gas. Adding any choice amount of dark matter will not solve the problem, as the increase in the infall velocity will be accompanied by an increase in the pre-collision sound speed of the gas, cancelling any potential increase in the Mach number of the collision, necessary to explain the observed shock.

The recent body of observational evidence reviewed her forces a change from considering GR plus dark matter, or modified gravity, as equivalently plausible options; in the low velocity, $a < a_0$ regime, gravity does not follow the standard descriptions of Einstein and Newton.

Acknowledgments Xavier Hernandez acknowledges financial assistance from UNAM DGAPA grant IN103011. Alejandra Jiménez acknowledges financial support from a CONACYT scholarship.

Bibliography

1. G.W. Angus, MNRAS. **387**, 1481 (2008)
2. C. Allen, E. Moreno, P. Pichardo, ApJ. **652**, 1150 (2006)
3. C. Allen, E. Moreno, P. Pichardo, ApJ. **674**, 237 (2008)
4. T. Bernal, S. Capozziello, J.C. Hidalgo, S. Mendoza, Eur. Phys. J. C. **71**, 1794 (2011)
5. J.D. Bekenstein, Phys. Rev. D. **70**, 083509 (2004)
6. J. Binney, S. Tremaine, *Galactic Dynamics*. (Princeton University Press, Princeton) (1987)
7. J.P. Bruneton, G. Esposito-Farese, Phys. Rev. D. **76**, 124012 (2007)
8. S. Capozziello, V.F. Cardone, A. Troisi, MNRAS. **375**, 1423 (2007)
9. S. Capozziello, M. De Laurentis,Phys. Rep. **509**, 167 (2011)

10. M.C. Chiu, C.M. Ko, Y. Tian, H.S. Zhao, Phys. Rev. D. **83**, 06352 (2011)
11. S. Dhital, A.A. West, K.G. Stassun, J.J. Bochanski, AJ. **139**, 2566 (2010)
12. G.A.Drukier, et al., AJ. **115**, 708 (1998)
13. P.A. Duc, et al., MNRAS. **417**, 863 (2011)
14. H. Haghi, H. Baumgardt, P. Kroupa, E.K. Grebel, M. Hilker, K. Jordi, MNRAS. **395**, 1549 (2009)
15. H. Haghi, H. Baumgardt, P. Kroupa, A&A. **527**, A33 (2011)
16. A. Halle, H.S. Zhao, B. Li, ApJS. **177**, 1 (2008)
17. X. Hernandez, S. Mendoza, T. Suarez, T. Bernal, A&A. **514**, A101 (2010)
18. X. Hernandez, M.A. Jiménez, C. Allen, EPJC. **72**, 1884 (2012a)
19. X. Hernandez, M.A. Jiménez, C. Allen, arXiv. 1206.5024 (2012b)
20. X. Hernandez, M.A. Jiménez, ApJ. **750**, 9 (2012)
21. Y.F. Jiang, S. Tremaine, MNRAS. **401**, 977 (2010)
22. P. Kroupa, arXiv. 1204.2546 (2012)
23. R.R. Lane, L.L. Kiss, G.F. Lewis, R.A. Ibata, A. Siebert, T.R. Bedding, P. Székely, MNRAS. **400**, 917 (2009)
24. R.R. Lane et al., MNRAS. **406**, 2732 (2010a)
25. R.R. Lane, L.L. Kiss, G.F. Lewis, R.A. Ibata, A. Siebert, T.R. Bedding, P. Székely, MNRAS. **401**, 2521 (2010b)
26. R.R. Lane et al., A&A. **530**, A31 (2011)
27. R.R. Lane, A.H.W. Küpper, D.C. Heggie, MNRAS. **423**, 2845 (2012)
28. J. Lee, E. Komatsu, ApJ. **718**, 60 (2010)
29. P.D. Mannheim, D. Kazanas, ApJ. **342**, 635 (1989)
30. C. Mastropietro, A. Burkert, MNRAS. **389**, 967 (2008)
31. S.S. McGaugh, Phys. Rev. Lett. **106**, 121303 (2011)
32. D.E. McLaughlin, R.P. Marel van der, ApJS. **161**, 304 (2005)
33. S. Mendoza, X. Hernandez, J.C. Hidalgo, T. Bernal, MNRAS. **411**, 226 (2011)
34. S. Mendoza, T. Bernal, X. Hernandez, J.C. Hidalgo, L.A. Torres, arXiv. 1208.6241 (2012)
35. J.W. Moffat, V.T. Toth, ApJ. **680**, 1158 (2008)
36. M. Milgrom, ApJ. **270**, 365 (1983)
37. M. Milgrom, ApJ. **429**, 540 (1994)
38. M. Milgrom, R.H. Sanders, ApJ. **678**, 131 (2008)
39. M. Milgrom, Phys. Rev. D. **82**, 043523 (2010)
40. N.R. Napolitano, S. Capozziello, A.J. Romanowsky, M. Capaccioli, C. Tortora, ApJ. **748**, 87 (2012)
41. R.H. Sanders, S.S. McGaugh, ARA&A. **40**, 263 (2002)
42. R. Scarpa, G. Marconi, R. Gilmozzi, in (*Proceedings of Baryons in Dark Matter Halos. Novigrad, Croatia, 5–9 Oct 2004*), Ed. by R. Dettmar, U. Klein, P. Salucci (SISSA, Proceedings of Science), p. 55.1, http://pos.sissa.it (2004)
43. R. Scarpa, G. Marconi, R. Gimuzzi, G. Carraro, A&A. **462**, L9 (2007a)
44. R. Scarpa, G. Marconi, R. Gimuzzi, G. Carraro, The Messenger. **128**, 41 (2007b)
45. R. Scarpa, R. Falomo, A&A. **523**, 43 (2010)
46. R. Scarpa, G. Marconi, G.Carraro, R. Falomo, S. Villanova, A&A. **525**, A148 (2011)
47. E.J. Shaya, R.P. Olling, ApJS. **192**, 2 (2011)
48. C. Skordis, D.F. Mota, P.G. Ferreira, C. Boehm, Phys. Rev. Lett. **96**, 011301 (2006)
49. Y. Sobouti, A&A. **464**, 921 (2007)
50. A. Sollima, C. Nipoti, MNRAS. **401**, 131 (2010)
51. R. Thompson, K. Nagamine. MNRAS. **419**, 3560 (2012)
52. S.C. Trager, I.R. King, S. Djorgovski, AJ. 109, 218 (1995)
53. H.S. Zhao, B. Famaey, Phys. Rev. D. **81**, 087304 (2010)

Part II
Cosmology and Dark Energy

Chapter 5
Vector Fields Resembling Dark Energy

Nora Bretón

Abstract We review how vector fields have been introduced to produce inflationary scenarios in early universes and recently they have been invoked to mimick dark energy. These last approaches have been mostly qualitatives, requiring then to be tested with cosmological probes, in order to seriously be considered as one of the possible causes of the present accelerated expansion of the universe.

5.1 Introduction

The recently obtained data from the observable universe constitute a breakthrough in cosmology, in the sense that it has become a theory that can be phenomenologically tested. Those observations point to an accelerated expanding universe, that is ruled by some kind of negative pressures ascribed to a dark entity. In the quest to explain the origin of this anti-gravity, a lot of work has been done, still lacking of a unanimous accepted explanation. There are several fields that are able to produce the negative pressure effect, among them are the vector fields and the nonlinear electromagnetic fields.

The need of producing an accelerated expanding universe is as old as the eighties, when inflation arose as a mechanism to solve the horizon and flatness problem. From the lessons learned to produce accelerated expansion in the early universe, we can try them also to explain the presently observed accelerated expansion.

In the next section we briefly review how vector fields were introduced to produce inflationary universes, following mainly the work by Ford [1], of a potential that depends on a vector field in a Bianchi I spacetime. Then we move on to explain why nonlinear electrodynamics can be considered as a good candidate to produce accelerated expansion, presenting two succesful examples of cosmologies with this type of field, that are singularity-free and inflationary.

Next we address the vector and nonlinear electromagnetic fields as candidates to substitute the cosmological constant that produces the nowadays cosmic acceleration. These models in spite of actually producing negative pressures, they have to be tested versus the cosmological evidence that we can access today; being the qualitative

---

Nora Bretón (✉)
Dpto. de Física, Centro de Investigación y de Estudios
Avanzados del I. P. N., Apdo. 14-740, México D.F., México,
e-mail: nora@fis.cinvestav.mx

C. Moreno González et al. (eds.), *Accelerated Cosmic Expansion*,
Astrophysics and Space Science Proceedings 38,
DOI 10.1007/978-3-319-02063-1_5,

behaviour not enough to rule out dark energy. In the last part of this contribution we show how to test a model versus cosmological probes, taking as particular case the non-Abelian Born-Infeld field coupled to a Friedmann–Lemaitre–Robertson–Walker (FLRW) spacetime. Turning out in the analyzed example that the accelerated expansion produced by the nonlinear electromagnetic component cannot substitute the dark energy component, to fit in a satisfactory way probes like the Supernovae Ia, Gamma-ray Bursts and Hubble parameter measurements.

5.2 Vector Fields Produce Inflation

To solve the problems of horizon and flatness, that is the explanation for the observed isotropy, it was proposed a scalar field (inflaton), whose dynamics governs the duration and end of inflation. The parameters of the inflaton must be finely tuned in order to allow adequate inflation and an acceptable magnitude for density perturbations.

L. Ford [1] discussed the possibility of inflationary models in which inflation is driven by a vector field. He assumed a vector field A^{μ} in a lagrangian of the form

$$L = \frac{1}{4} F_{\mu\nu} F^{\mu\nu} + V(\xi), \tag{5.1}$$

where $F_{\mu\nu} = \partial_{\mu} A_{\nu} - \partial_{\nu} A_{\mu}$ is the electromagnetic field tensor. The vector field is self-coupled through the potential $V(\xi = A_{\rho} A^{\rho})$; indeed V can be considered as an effective interaction arising from the coupling of A_{μ} to other matter fields. The interaction might arise from a gauge field with spontaneously broken $U(1)$ symmetry, much as the massive vector field arises in the Abelian-Higgs model. Assuming a potential of the form $A_{\mu} = \delta^{z}_{\mu} A(t)$ the corresponding stress-energy tensor is not isotropic, then the spacetime to be coupled to cannot be of the isotropic Robertson-Walker (RW) type. Instead, this field can be coupled to an anisotropic metric like Bianchi I,

$$ds^2 = -dt^2 + a^2(t)(dx^2 + dy^2) + b^2(t)dz^2, \tag{5.2}$$

where $a(t)$, $b(t)$ are the expansion factors that are not the same in all directions. Actually such a potential V can be found that the universe expands acceleratedly and then approaches a de Sitter space. As long as the form of the potential presents a flat region then there is a period of isotropic inflation (see Fig. 5.1). After inflation ends, anisotropy could be developed; however if the anisotropy ceases before the CMB is formed, no conflicts with observation will arise. The isotropization problem can be avoided if a colissionless fluid is present, capable of canceling the anisotropy in the vector-field stress tensor, i.e. particles produced at the reheating time but which subsequently do not interact significantly apart from their gravitational effects. If this cancelation of anisotropies occurs, it results a radiation-dominated RW universe.

A reheating period is a requirement of a good inflationary model: In this model gravitational particle creation will be enhanced by the anisotropy and hence become a more efficient reheating mechanism.

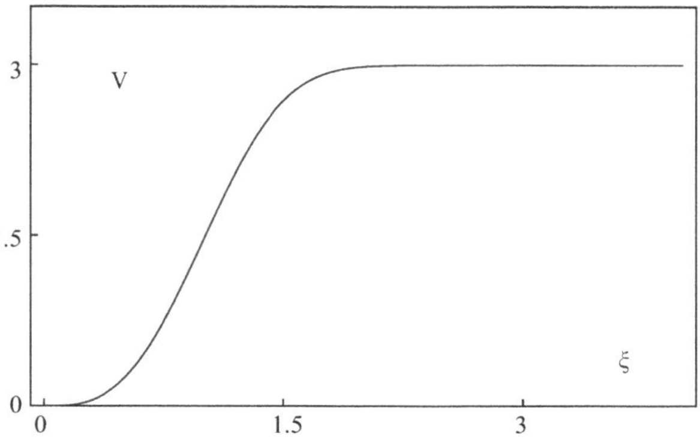

Fig. 5.1 Such potential with a plateau and a reheating zone in fact can be found as function of ξ. The plot is for $V(\xi) = V_0(1 - e^{-m^2\xi/2V_0})$, the reheating zone is near $\xi = 0$

In summary, vector fields can drive anisotropic inflationary expansion that is still capable of solving the same cosmological puzzles (horizon and flatness) as isotropic inflation. Suitable choices of the values of the parameters of the model can be found and the exceding anisotropy at late times may be avoided. Moreover, the massless vector field is conformally invariant whereas the massless minimally coupled scalar field is not. Ford also suggested that vector fields would also produce a scale invariant perturbation spectrum. The squeme so far is pretty much like quintessence.

5.3 Nonlinear Electrodynamics

As early as the 1930's, it was found that at high energies electromagnetic phenomena should not obey Maxwell electrodynamics, and nonlinear lagrangians we proposed to account for nonlinear electromagnetic effects. Such effects like photon-photon interaction or the creation and subsequent annihilation of electron-positron pairs are nonlinear effects that take place at high energies of the order of the critical fields of Schwinger, $E \approx 10^{18}$ Volt/m, or $B \approx 10^{13}$ Gauss; such scenario is very reasonable to expect in early universes. Therefore we can also try vector fields in nonlinear lagrangians L_{NLEM} coupled to gravity, to produce inflation, with actions of the form,

$$S = \int d^4x\sqrt{-g}\{\frac{R}{16\pi} - L_{\mathrm{NLEM}}\}, \tag{5.3}$$

where R denotes the scalar curvature, $g := \det|g_{\mu\nu}|$ and L_{NLEM} is the electromagnetic part, that depends in nonlinear way on the invariants of the electromagnetic field, F, G,

$$F = F_{\mu\nu}F^{\mu\nu}, \quad G = \tilde{F}_{\mu\nu}F^{\mu\nu}. \tag{5.4}$$

where $\tilde{F}_{\mu\nu}$ is the dual of the electromagnetic field tensor. Moreover, nonlinear electromagnetic fields are characterized by some desirable inflation-like features. The nonvanishing of the trace gives rise to the breakdown of conformal invariance. This effect, also known as conformal or Weyl anomaly, may arise by the interaction of photons with other fields. During inflation it provides a mechanism to amplify perturbations, like primordial magnetic fields, for instance, the seed magnetic fields of the nowadays observed magnetic fields in galaxies [2–4].

Another appealing feature of NLEM is the possibility of producing negative pressures as a consequence of the violation of the strong energy condition (SEC). SEC can be posed as

$$R_{\mu\nu}V^{\mu}V^{\nu} \geq 0, \tag{5.5}$$

and using the Einstein equations can be settled as,

$$R_{\mu\nu}V^{\mu}V^{\nu} = 8\pi\left(T_{\mu\nu}V^{\mu}V^{\nu} + \frac{T}{2}\right) \geq 0. \tag{5.6}$$

that for an isotropic perfect fluid with traceless $T_{\mu\nu}$ amounts to $\rho + 3\,p \geq 0$. Note that any matter tensor $T_{\mu\nu}$ whose trace is negative enough, can violate SEC. In particular, nonlinear electromagnetic lagrangians, $L(F, G)$, with energy-momentum tensor and trace given by

$$\begin{aligned} 4\pi T_{\mu\nu} &= -L_{,F}F_{\mu}^{\alpha}F_{\alpha\nu} + (GL_{,G} - L)g_{\mu\nu}, \\ -8\pi T &= 8\pi(L - FL_{,F} - GL_{,G}) = R. \end{aligned}$$

may not hold SEC but still fulfil the dominant energy condition (DEC), namely, $T_{\mu\nu}V^{\mu}V^{\nu} > 0$. NLEM has been also useful in preventing the ocurrence of the initial singularity is some cosmological models. The reason can be traced from the Raychaudhuri equation. Let us consider a time like congruence $\Theta(\lambda)$, whose rate of change of expansion is governed by the Raychaudhuri equation,

$$\frac{d\Theta(\lambda)}{d\lambda} = -R_{ab}V^{a}V^{b} + 2\omega^2 - 2\sigma^2 - \frac{\Theta^2}{3} + \dot{V}^{\alpha}_{;\alpha}, \tag{5.7}$$

where ω is the twist, σ is the shear, and $\dot{V}^{\alpha}$ is the acceleration of the congruence.

If $\frac{d\Theta(\lambda)}{d\lambda} < 0$ then the congruence tends to converge and a singularity arises. Otherwise, if $\frac{d\Theta(\lambda)}{d\lambda} > 0$ the focusing does not occur. Let us consider a geodesic congruence, $\dot{V}^{\alpha} = 0$. If $R_{ab}V^{a}V^{b} < 0$ (violation of SEC) and $-R_{ab}V^{a}V^{b} > -2\omega^2 + 2\sigma^2 + \frac{\Theta^2}{3}$, then the possibility exists of avoiding singularities by stopping the focusing of geodesics.

These NLEM features have been implemented in some early universe models. In [5] it was proposed a G_2 inhomogeneous spacetime, conformal to Bianchi metrics G_3IX, G_3II, G_3III, G_3VIII as special cases, coupled to a Born-Infeld electromagnetic field. As was expected, a nonsingular, accelerated expanding cosmology was found, whose anisotropy vanish at large times. The NLEM radiation that is governed by the state equation $p = -\frac{\rho}{\rho+1}$, for $\rho \gg 1$ approaches -1 [6]. The shear, σ, that measures anisotropy, vanishes as time goes by; the deceleration parameter, q, being

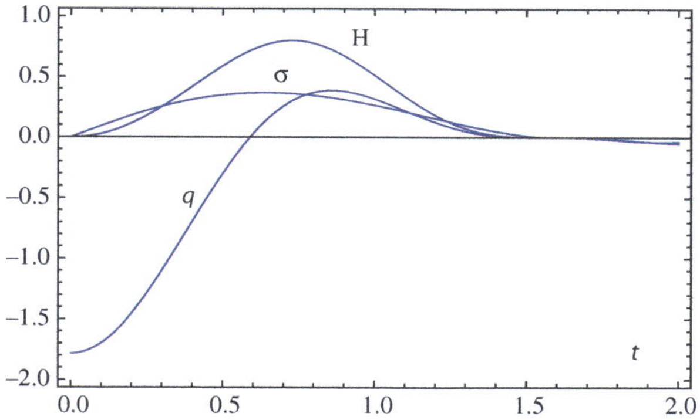

Fig. 5.2 Qualitative behaviour for the BI cosmology [6], of the mean Hubble parameter, H, the shear, σ, and the deceleration parameter, q, that shows the accelerated expansion

negative during a period of time, indicating the accelerated expansion; these features are shown in Fig. 5.2.

Another example is bouncing cosmology [7, 8], where the initial singularity is avoided using a lagrangian of the form

$$L = -\frac{F}{4} + \alpha F^2. \tag{5.8}$$

In a RW geometry, this field produced the effect that the scale factor approaches zero but never reaches it, instead it has a bouncing and then expands acceleratedly (see Fig. 5.3). Other similar models of NLEM in FLRW were proposed later [9].

The way of implementing inflationary scenarios gave us a lesson of what kind of actions or lagrangians can produce accelerated expansions. We now try to implement some of those recipes to mimick dark energy with vector fields.

5.4 Vector Fields Mimicking Dark Energy

Assuming homogeneous and isotropic geometry, as observations led us to conclude, it is customary to model the cosmic substratum as a perfect fluid, this is the Friedmann-Lemaitre-Robertson-Walker cosmological model. In this model there is only one function to be determined, the scale factor $a(t)$, provided we have input the equation of state (EoS) of the fluid. Under these assumptions, Einstein equations are

$$\left(\frac{\dot{a}}{a}\right)^2 = \frac{8\pi G}{3c^2}\rho, \quad \left(\frac{\ddot{a}}{a}\right) = -\frac{4\pi G}{3c^2}(\rho + 3\, p), \tag{5.9}$$

From the second equation it is clear that to have an accelerated universe, $\ddot{a}(t) > 0$, we need negative pressures, $(\rho + 3\, p) < 0$. This implies the violation of SEC.

As we learned from the inflation problem, vector fields can do the job and to this purpose, some vector-tensor theories of gravity have been proposed. In [10], actions of self-coupling vector fields, and vector fields coupled to gravity have been

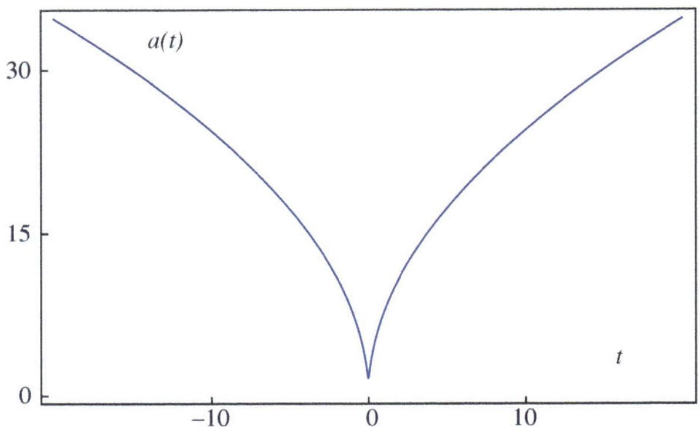

Fig. 5.3 The singularity-free bouncing cosmology. Qualitative plot of the scale factor as function of t, $a^2 = H_0\sqrt{2t^2/3 + 8\alpha}$; $a(t)$ never reaches $t = 0$

considered, as compelling candidates to drive an accelerated phase. Moreover, the authors found six models that fulfil that their PPN parameters behave like GR ones.

In [11] it is considered an action depending on the electromagnetic field, without assuming the Lorenz gauge, $\nabla_\mu A^\mu = 0$,

$$S = \int d^4x\sqrt{-g}\left\{-\frac{1}{4}F_{\mu\nu}F^{\mu\nu} + \frac{\xi}{2}(\nabla_\mu A^\mu)^2 + A_\mu J^\mu\right\}, \tag{5.10}$$

with no new scales introduced to the action. For systems whose dynamics derives from this action it has been shown that attractors with accelerated expansion can give rise to dark energy candidates. Moreover, when a matter component is introduced in addition to the vector fields, then there are solutions with a transition from a matter dominated universe to a phase of accelerated expansion. $(\nabla_\mu A^\mu)$ behaves as a free scalar field non-minimally coupled to gravity and in an expanding universe, it gets frozen for super Hubble modes, so that the gauge fixing term in Eq. (5.10) shall give an effective cosmological constant on large scales.

5.5 Nonlinear Electromagnetism as Dark Energy

Again, NLEM can be invoked as a field that can drive an accelerated phase of the universe. One of the most interesting properties of NLEM is the breakdown of conformal invariance that leads to the occurrence of negative pressures. This breakdown of conformal symmetry can be originated from quantum corrections, non-minimal coupling to gravity, dilaton or other coupled scalar fields or string theory corrections. On this line there are several proposals [12, 13].

Novello et al. [14] proposed a complete or unified scenario to describe the universe, from the early inflation to the present acceleration, avoiding the initial singularity, with parameters that should be adjusted from confrontation to observations, the considered action was,

$$S_{NLEM} = \int \sqrt{-g}d^4x\left(\alpha F^2 - \frac{F}{4} + \frac{\gamma}{F}\right) \tag{5.11}$$

In string theory it stands out the Born-Infeld action [15]. The Born-Infeld theory was proposed in the 1930's, with the aim to avoid the self-energy divergence of the electron field [16]; characterized by a maximal field strenght β, its action is given by

$$S_{BI} = \int \sqrt{-g} d^4 x \beta^2 \left(1 - \sqrt{1 - \frac{F^2}{\beta^2} + \frac{G^2}{\beta^4}}\right) \tag{5.12}$$

This theory applied to FLRW cosmology have been addressed by several authors, for instance, Elizalde et al. [17], considered a BI isotropic cosmology where quantum condensates appear due to vacuum fluctuations. Dyadichev et al. studied quantum condensates in a Yang-Mills cosmology [18].

5.6 Electromagnetic Fields in Isotropic Spaces: A Technical Problem

So far we have not mentioned the technical difficulty that arises when one tries to couple vector fields to isotropic RW geometry, because vector fields generically single out spatial directions, leading to anisotropy.

To solve this problem, a spatial average in the fields must be taken, in order to obtain an isotropic energy momentum tensor. Several authors adopt the proposal of spatial average by Tolman and Ehrenfest (1933) [19]; they considered the thermal equilibrium of a general static gravitational field which could correspond to a system containing solid as well as fluid parts. Besides, they take a proper reference system in which the radiation as a whole is at rest. Vollick [20] justifies the average as follows: Spacetime is filled with electromagnetic radiation, the one that is of cosmological interest is that of the CMB. It can be considered as a stochastic background of short wavelength radiation (compared with curvature) that satisfies the spatial average for volumes that are large compared to the wavelength but small compared to the curvature ($R \gg$ volume $\gg \lambda_{\mathrm{CMB}}$).

5.6.1 The Spatial Average

Defining the volumetric spatial average of X at the time t by

$$\langle X \rangle \equiv \lim_{V \to V_0} \frac{1}{V} \int X \sqrt{-g} d^3 x, \tag{5.13}$$

By identifying the electric and magnetic components as

$$E_i = F_{i0}, \quad B_i = \frac{1}{2} \epsilon_{ijk} F^{jk}, \tag{5.14}$$

the spatial average leads to

$$\langle F^{0}_{.\rho} F^{0\rho} \rangle = E^2, \quad \left\langle F_{i\rho} F^{\rho}_{j} \right\rangle = -\langle E_i E_j \rangle + 2\langle B_i B_j \rangle, \tag{5.15}$$

Then for an observer who finds on the average no net flow of energy in the radiation field,

$$\begin{aligned} \langle E_i B_j \rangle &= 0, \\ \langle E_i E_j \rangle = \frac{g_{ij}}{3} E^2, &\quad \langle B_i B_j \rangle = \frac{g_{ij}}{3} B^2, \\ \langle E_i \rangle = \langle B_i \rangle = 0, &\quad \langle E_i B_j \rangle = 0, \end{aligned} \tag{5.16}$$

besides, advocating the equipartition principle, $E^2 = B^2$ the process ends up with an isotropic stress-energy tensor.

5.6.2 *The Cosmic Triad*

Alternatively, to couple electromagnetic fields to isotropic metrics, it has been considered a set of three self-interacting vector fields A^a_μ, known as the *cosmic triad* [21]. The cosmic triad can arise from a non-Abelian SU(2) gauge theory, like Yang-Mills, assuming gauge fields as functions only of time, with

$$\begin{aligned} F^a_{\mu\nu} &= \partial_\mu A^a_\nu - \partial_\nu A^a_\mu + \epsilon^a_{bc} A^b_\mu A^c_\nu, \\ A^b_\nu &= \delta^b_\nu A(t) a, \end{aligned} \tag{5.17}$$

where ϵ^a_{bc} is the structure constant of the gauge group SU(2). Then $E_1^2 = E_2^2 = E_3^2$, $B_1^2 = B_2^2 = B_3^2$, and $F = E^2 - B^2 = \sum_{i=1}^3 E_i^2 - \sum_{i=1}^3 B_i^2$, such that $T_{\mu\nu} = \sum_{a=1}^3 E_i^{2(a)} T_{\mu\nu}$ has isotropic stresses. This is the so called *triad*, that is a triad in an internal space at each point of the spacetime.

After taking some of these averages, the result is an isotropic stress-energy tensor that in terms of a general nonlinear lagrangian $L(F, G)$ is given by,

$$T_{\mu\nu} = -4 L_F F^{\alpha}_{\mu.} F_{\alpha\nu} + (G L_G - L) g_{\mu\nu}, \tag{5.18}$$

or in terms of the fluid energy density ρ and pressure p,

$$\begin{aligned} T_{\mu\nu} &= (\rho + p) u_\mu u_\nu - p g_{\mu\nu}, \\ \rho &= -L + G L_G - 4E^2 L_F, \\ p &= L - G L_G + \frac{4}{3}(E^2 - 2B^2) L_F, \end{aligned} \tag{5.19}$$

5.7 Observational Constraints on Non-Abelian Born-Infeld Model

We shall address now the non-Abelian electromagnetic field with a Born-Infeld lagrangian. Considering the Non-Abelian Born-Infeld (NABI) Lagrangian

$$L = \left\{ 1 - \sqrt{1 - \frac{F^a_{\mu\nu} F_a^{\mu\nu}}{2\beta^2} + \frac{(\tilde{F}^a_{\mu\nu} F_a^{\mu\nu})^2}{16\beta^4}} \right\} \beta^2 \tag{5.20}$$

β being a *critical field* to be determined from observations. The cosmic triad can be implemented with a function, $w(t)$. In the RW geometry, with scale factor $a(t)$, the coupled Einstein-Yang-Mills (EYM) equations can be solved, the corresponding negative pressure being still insufficient to mimic inflation, with a EoS given by

$$p = \frac{\rho\,(\rho_c - \rho)}{3\,(\rho_c + \rho)}, \tag{5.21}$$

that interpolates between $p = -\rho/3$ (gas of Nambu-Goto strings) and $p = \rho/3$ (radiation) at large time. Near the singularity the oscillations of the YM field are damped. In [22] was presented the dynamical behaviour of this system; in particular, for flat spacetime, $k = 0$, there is a singularity at the origin. The coupled equations for $w(t)$ and $a(t)$ can be integrated,

$$\frac{\dot{w}^2}{(a_0^4 - w^4)} = \frac{a^2}{(a^4 + 3a_0^4)}, \tag{5.22}$$

the analytical expression for $w(t)$ is in terms of Jacobi elliptic functions. The generic solution for $w(t)$ is oscilatory. For small $a(t)$, near the singualirity, $a \to 0$, the YM oscillations slow down in the NABI case, this is shown in Fig. 5.4.

Considering dark matter Ω_m, dark energy Ω_Λ, and the NABI contribution, Ω_g, as the components of the universe, the Hubble function is given by

$$\frac{H(z)^2}{H_0^2} = \Omega_m(1+z)^3 + (1 - \Omega_m - 2\Omega_g) + \Omega_g\sqrt{1 + 3(1+z)^4}, \tag{5.23}$$

with $\Omega_i = \frac{8\pi}{3\,H^2}\rho_i$, Ω_g is the abundance of the radiation with NABI behaviour.

We present a preliminar test of the NABI model with a chi-square statistics, minimazing the χ^2,

$$\chi^2 = \left(\frac{A_{th} - A_{Obs}}{\sigma_A}\right)^2, \tag{5.24}$$

for the following probes: SNIa Union 2.1 [23], observational Hubble parameter [24] and GRBs [25, 26]. The results of the statistical analysis are shown in Figs. 5.5–5.8. In Fig. 5.5 it is shown the confidence contours for the three probes: the best fit for SNe Ia and Hubble parameter agree with the Λ CDM model in the sense that the

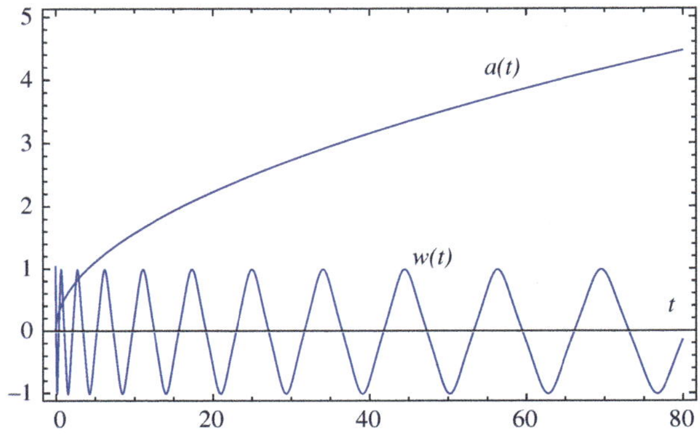

Fig. 5.4 Behaviour of the scale factor $a(t)$ and the gauge field function $w(t)$ for the NonAbelian Born-Infeld spatially flat universe [22]

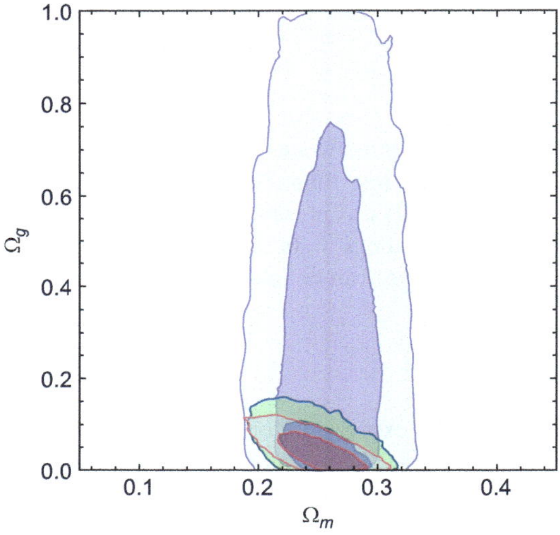

Fig. 5.5 Confidence contour for NABI model tested for the parameters (Ω_g, Ω_m) for three cosmological probes: SNe Ia, Hubble parameter and GRBs

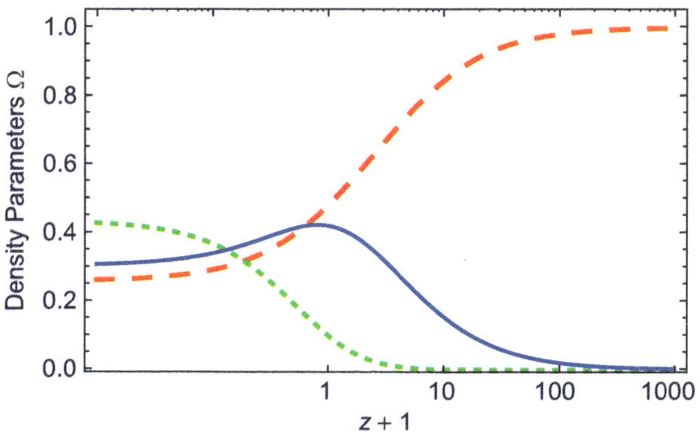

Fig. 5.6 The curves of abundance of the three components considered: Ω_Λ (small dashed), Ω_g (continuous), Ω_m (long dashed), for NABI model considering only the GRBs probe

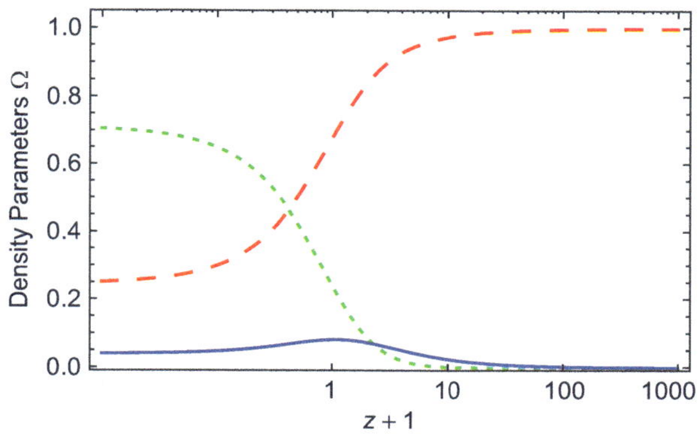

Fig. 5.7 Abundance of the three components considered: Ω_Λ (small dashed), Ω_g (continuous), Ω_m (long dashed), for NABI model considering three probes: GRBs, Hubble measurements and SNe Ia

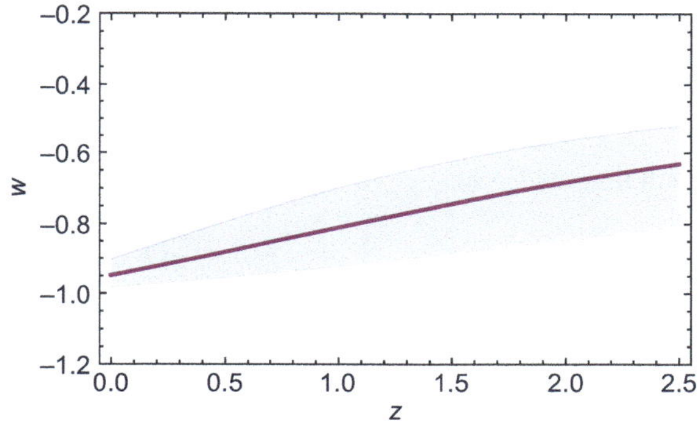

Fig. 5.8 The evolution of the parameter of the EoS of the dark energy for the NABI cosmology, $p_{de} = w(z)\rho_{de}$, obtained from the joint three probes. It clearly approaches the value -1 at present time

BI component is not very relevant, being ($\Omega_g = 0.048, \Omega_m = 0.25$) and ($\Omega_g = 0.048, \Omega_m = 0.25$), respectively. However, for the wider range of the GRBs data, $1.4 < z < 8.1$, the BI component is enhanced being ($\Omega_g = 0.3021, \Omega_m = 0.26$); the increment in Ω_g occurs at the expense of Ω_Λ, diminishing this later to $\Omega_\Lambda \approx 0.44$. Unfortunately the confidence in GRBs data is not as good as the two first probes, SNe Ia and Hubble parameter. The previous results are illustrated in terms of the component abundances in Figs. 5.6 and 5.7. Finally Fig. 5.8 illustrates the evoultion of the parameter of the equation of state of dark energy ($p_{\rm de} = w(z)\rho_{\rm de}$), $w(z)$, as a function of the redshift; the tendency as time goes by is to reach -1, approaching from the left, $w(z) > -1$. The analysis of several Born-Infeld cosmologies tested versus cosmological probes can be consulted in [27]

5.8 Conclusions

Vector fields and nonlinear electromagnetic fields can produce accelerated expansion and other desirable features in cosmology, i.e. they give good qualitative phenomenology. Nonlinear electromagnetic fields are very reasonable to assume for inflationary epochs (high energies, anisotropy). However in late universes it is not well understood what its origin could be, some authors have relied the NLEM presence to the occurrence of a sort of quantum condensates, being the conformal symmetry breaking an appealing NLEM feature.

To discard dark energy in favor of a vector or tensor field, it should not only accelerate the universe but do it at the right time and at the right amount, according to observational data. When probed with GRBs, the NABI cosmology shows that NLEM contribution can be important at intermediate redshifts, not completely replacing the whole dark energy, but a good proportion. However the dispersion in GRBs data still is not reliable.

The negative pressure originated in the examined case seems to be insufficient to mimic dark energy. However, if we consider a $L(F, G)$ completely general, after isotropy is implemented, the obtained isotropic fluid is characterized by

$$\rho = -L + GL_G - 4L_F E^2,$$
$$p = L - GL_G - \frac{4}{3}(2B^2 - E^2)L_F, \tag{5.25}$$

such that the EoS is

$$p = -\rho - \frac{8}{3}(E^2 + B^2)L_F, \tag{5.26}$$

that is, negative pressures can arise from such a fluid, so the possibility exist that an electromagnetic field can account for dark energy or at least to diminish its necessity.

Acknowledgments N. B. acknowledges Claudia Moreno and the Organizing Committee (CUCEI-UdeG) of the IV Int. Meeting on Gravitation and Cosmology for the invitation to lecture.

References

1. L.H. Ford, Inflation driven by a vector field. Phys. Rev. D **40**, 967 (1989)
2. K.E. Kunze, Primordial magnetic fields and nonlinear electrodynamics. Phys. Rev. D **77**, 023530 (2008)
3. H.J. Mosquera Cuesta, G. Lambiase, Primordial magnetic fields and gravitational baryogenesis in nonlinear electrodynamics. Phys. Rev. D **80**, 023013 (2009)
4. L. Campanelli, P. Cea, G.L. Fogli, Inflation produced magnetic fields in nonlinear electrodynamics. Phys. Rev. D **77**, 043001 (2008)
5. R. García-Salcedo, N.Bretón, Born-Infeld cosmologies. Int. J. of Modern Physics A **15**, 4341–4353 (2000)
6. R. García-Salcedo, N. Bretón, Nonlinear electrodynamics in Bianchi spacetimes. Class. Quantum Grav. **20**, 5425–5437 (2003)
7. M. Novello, S.E. Perez Bergliaffa, Bouncing cosmologies. Phys. Rep. **463**, 127–213 (2008)
8. M. Novello, E. Goulart, J.M. Salim, S.E. Perez Bergliaffa, Cosmological effects of nonlinear electrodynamics. Class. Quantum Grav. **24**, 3021 (2007)
9. C.S. Camara, M.R. deGarcia Maia, J.C. Carvalho, J.A.S. Lima, Nonsingular FRW cosmology and nonlinear electrodynamics. Phys. Rev. D **69**, 123504 (2004)

10. J. Beltrán Jiménez, A.L. Maroto, Avoiding the dark energy coincidence problem with a cosmic vector. AIP Conf. Proc. **1122**, 107–114 (2009) Physics and Mathematics of Gravitation: Proceedings of the Spanish Relativity Meeting (2008)
11. J. Beltrán Jiménez, L. A. Maroto, Electromagnetic nature of dark energy. J. Cosmol. Astropart. JCAP **03**, 016 (2009)
12. L. Labun, J. Rafelski, Dark energy simulacrum in nonlinear electrodynamics. Phys. Rev. D **81**, 065026 (2010)
13. E. Mottola, in *Conformal Invariance, Dynamical Dark Energy and the CMB*, (2011) Proceedings of the XLVth Rencontres de Moriond, 2010 Cosmology, ed. by E. Auge, J. Dumarchez, J. Tran Tranh Van, (Vietnam, The Gioi Publishers) (2010) [arXiv:1103.1613].
14. M. Novello, S.E. Perez Bergliaffa, J.M. Salim, Nonlinear electrodynamics and the acceleration of the universe, Phys. Rev. D, **69**, 127301 (2004)
15. E. Bergshoeff, E. Sezgin, C.N. Pope, P.K. Townsend, The Born-Infeld action from conformal invariance of the open superstring. Phys. Lett. B **188**, 70 (1987)
16. M. Born, L. Infeld, Foundations of the new field theory. Proc. Roy. Soc. Lond. **A144**, 125 (1934).
17. E. Elizalde, J.E. Lidsey, S. Nojiri, S.D. Odintsov, Born-Infeld quantum condensate as dark energy in the universe. Phys. Lett. B **574**, 1 (2003)
18. D.V. Gal'tsov, M.S. Volkov, Yang-Mills cosmology: cold matter for a hot universe. Phys. Lett. B **256**, 17–21 (1991)
19. R.C. Tolman, P. Ehrenfest, Temperature Equilibrium in a static gravitational field. Phys. Rev. **36**, 1791 (1930)
20. D.N. Vollick, Homogeneous and isotropic cosmologies with nonlinear electromagnetic radiation. Phys. Rev. D **78**, 063524 (2008)
21. C. Armendáriz-Picón, Could dark energy be vector-like?. J. Cosm. Astrop. Phys. **JCAP07**, 007 (2004)
22. V.V. Dyadichev, D.V. Gal'tsov, and A.G. Zorin, M. Yu, Non-Abelian Born-Infeld cosmology. Phys. Rev. D **65**, 084007 (2002)
23. N. Suzuki et al., The Hubble Space Telescope Cluster Supernova Survey: V. Improving the Dark Energy Constraints Above $z > 1$ and Building an Early-Type-Hosted Supernova Sample [arXiv:1105.3470]
24. M. Moresco et al., New constraints on cosmological parameters and neutrino properties using the expansion rate of the Universe to z$\sim$1.75 [arXiv:1201.6658]
25. Y. Wang, Model-Independent Distance Measurements from Gamma-Ray Bursts and Constraints on Dark Energy. Phys. Rev. D **78**, (2008) 123532 [arXiv:0809.0657]
26. H. Wei, Observational constraints on cosmological models with the updated long gamma-ray burst. JCAP **08**, 020 (2010) [arXiv:1004.4951]
27. N. Bretón, R. Lazkoz, A. Montiel, Observational constraints on electromagnetic Born- Infeld cosmologies. JCAP **10**, 013 (2012)

Chapter 6
Diameter Angular Distance in Locally Inhomogeneous Models

Carlos Custodio and Nora Bretón

Abstract Locally inhomogeneous cosmological models have been proposed since the sixties. Recently these models have been revisited in order to explain the observed accelerated expansion of the universe. Application of the optical scalar equations in such models has given estimates of the distortions in the observations of distant objects and the distance-redshift relations. In this work we analyze the influence of local inhomogeneities on the trajectory of light beams, testing the ZKDR luminosity distance with three cosmological probes: supernovae Ia, gamma-ray bursts and Hubble parameter measurements, in a wide redshift range of $0.1 < z < 8.1$.

6.1 Introduction

Recent observations of the type Ia supernovae and CMB anisotropy strongly indicate that the total matter-energy density of the universe is now dominated by some kind of vacuum energy, the so called "dark energy" or the cosmological constant Λ [1]. The origin and nature of this vacuum energy remain unknown. On the other hand, it is well known from galaxy surveys that galaxies and clusters of galaxies up to a scale of $\simeq$ 1Gpc are distributed non homogeneously forming filaments, walls and underdense voids. This indicates that on similar scales also the dark matter is distributed non homogeneously. In this work we analyze the effect of local inhomogeneities on the angular diameter distance during the propagation of a light beam in a universe with non zero cosmological constant. We test the Zeldovich–Kantowski–Dyer–Roeder (ZKDR) distance with observations of Gamma-Ray Bursts (GRBs) in a range of redshift of $1.4 < z < 8.1$. To make our analysis more robust, we include the Supernova and Hubble parameter probes, in the redshift range of $0.1 < z < 1.7$.

C. Custodio (✉) · N. Bretón
Dpto. de Física, Centro de Investigación y de Estudios Avanzados del I. P. N., Apdo. 14-740, D.F., México
e-mail: cromero@fisica.ufpb.br

Nora Bretón
e-mail: nora@fis.cinvestav.mx

C. Moreno González et al. (eds.), *Accelerated Cosmic Expansion*, Astrophysics and Space Science Proceedings 38,
DOI 10.1007/978-3-319-02063-1_6,

6.2 General Considerations

Let us consider a beam of light emanating from a source S in an arbitrary spacetime described by the metric tensor $g_{\alpha\beta}$. The light rays propagate along a null surface Σ, which is determined by the eikonal equation,

$$g^{\alpha\beta}\Sigma_{,\alpha}\Sigma_{,\beta} = 0. \tag{6.1}$$

The vector field tangent to the light ray congruence with the affine parameter λ, $k^\alpha = \frac{dx^\alpha}{d\lambda} = -\Sigma_{,\alpha}$, determines two optical scalars, the expansion θ and the shear σ, which are defined by

$$\theta = \frac{1}{2}k^\alpha_{;\alpha}, \quad \sigma = k_{\alpha;\beta}\overline{m}^\alpha\overline{m}^\beta. \tag{6.2}$$

The optical scalars θ and σ describe the relative rate of change of an infinitesimal area A of the cross section of the beam of light rays and its distortion. These two optical scalars satisfy the Sachs [2] propagation equations

$$\dot{\theta} + \theta^2 + \sigma^2 = -\frac{1}{2}R_{\alpha\beta}k^\alpha k^\beta, \tag{6.3}$$

$$\dot{\sigma} + 2\theta\sigma = -\frac{1}{2}C_{\alpha\beta\gamma\delta}\overline{m}^\alpha k^\beta \overline{m}^\gamma k^\delta. \tag{6.4}$$

Since we shall address a flat Robertson-Walker (RW) geometry, the Weyl tensor, $C_{\alpha\beta\gamma\delta}$, is zero and the shear vanishes, $\sigma = 0$, then we are left only with Eq. (6.3).

The expansion θ is related to the relative change of an infinitesimal surface area A of the beam's cross section by

$$\theta = \frac{1}{2}\frac{d\ln A}{d\lambda}. \tag{6.5}$$

Considering that the angular diameter distance D_A is proportional to $\sqrt{A}$, using D_A instead of $\sqrt{A}$ and replacing the affine parameter λ by the redshift z, then Eq. (6.3) can be written as [3],

$$\left(\frac{dz}{d\lambda}\right)^2\frac{d^2D_A}{dz^2} + \left(\frac{d^2z}{d\lambda^2}\right)\frac{dD_A}{dz} + \frac{4\pi G}{c^4}T_{\alpha\beta}k^\alpha k^\beta D_A = 0, \tag{6.6}$$

where we have substituted the Ricci tensor $R_{\alpha\beta}$ in terms of the energy-momentum tensor $T_{\alpha\beta}$ using the Einstein equations, $R_{\alpha\beta} - Rg_{\alpha\beta}/2 = 8\pi GT_{\alpha\beta}$. The appropriate initial conditions, such that the solid angle generated by the angular diameter distance satisfies that [area = radius^2 $\times$ solid angle], are

$$D_A(z)\mid_{z=0} = 0, \quad \frac{dD_A(z)}{dz}\mid_{z=0} = \frac{c}{H_0}, \tag{6.7}$$

where H_0 is the present value of the Hubble parameter and c is the velocity of light.

6.3 The ZKDR Luminosity Distance

To apply the angular diameter distance to the realistic universe it is necessary to take into account local inhomogeneities in the distribution of matter. Following Dyer & Roeder [4], we introduce a phenomenological parameter $\alpha = (1 - \rho_{\text{clumps}}/\rho)$ called the clumpiness parameter, which quantifies the amount of matter in clumps relative to the amount of matter distributed uniformly. The values of the parameter α for a partially clumped universe are in the range $0 < \alpha < 1$; being $\alpha = 1$ the FRW limit and $\alpha = 0$ corresponding to a totally clumped universe.

Therefore we shall consider an energy-momentum tensor $T_{\alpha\beta}$ given by,

$$T_{\alpha\beta} = \alpha\rho_m u_\alpha u_\beta + \rho_\Lambda g_{\alpha\beta}, \tag{6.8}$$

where u^α is the four-velocity of a comoving (with the cosmic fluid) observer, ρ_m is the matter energy density and ρ_Λ is the energy density of the cosmological constant.

We shall assume the homogeneous and isotropic geometry of Robertson-Walker (RW), with line element given in coordinates (t, r, θ, ϕ) by

$$ds^2 = c^2dt^2 - R^2(t)\left(dr^2 + r^2(d\theta^2 + \sin^2\theta d\phi^2)\right), \tag{6.9}$$

where we are considering zero curvature. From the Einstein equations we obtain that the Hubble parameter $H(z)$, defined as the rate of expansion, is given by,

$$H^2(z) = \left(\frac{\dot{R}}{R}\right)^2 = \sum_i \frac{8\pi G\rho_i}{3H_0^2c^2}, \tag{6.10}$$

where the sum is over the i-matter components considered in the model. Besides, defining the fractional densities of the ρ_i matter component of the universe by

$$\Omega_i = \frac{8\pi G}{3H_0^2c^2}\rho_i, \tag{6.11}$$

Eq. (6.6) can be rewritten in the form

$$\left(\frac{dz}{d\lambda}\right)^2 \frac{d^2D_A}{dz^2} + \left(\frac{d^2z}{d\lambda^2}\right)\frac{dD_A}{dz} + \frac{3}{2}\alpha\Omega_m(1+z)^5 D_A = 0. \tag{6.12}$$

Considering only two components in the universe: the dark matter (dust), ρ_m, and cosmological constant energy, ρ_Λ, the Hubble parameter is given by

$$H(z) = H_0\sqrt{\Omega_m(1+z)^3 + \Omega_\Lambda}. \tag{6.13}$$

Transforming from the affine parameter to the observable redshift, using that $\frac{dz}{d\lambda} = (1+z)^2 H(z)/H_0$, into Eq. (6.12) we obtain

$$(1+z)[\Omega_m(1+z)^3 + \Omega_\Lambda]\frac{d^2 D_A}{dz^2} + \frac{3}{2}\alpha\Omega_m(1+z)^2 D_A$$
$$+\left(\frac{7}{2}\Omega_m(1+z)^3 + 2\Omega_\Lambda\right)\frac{D_A}{dz} = 0. \tag{6.14}$$

The general properties of Eq. (6.14) have been extensively studied by Kantowski [5, 6]. With the appropriate change of variable, Eq. (6.14) can be transformed to a Legendre equation, whose solutions are the Legendre functions. Equation (6.14) can also be transformed into a hypergeometric equation, with the appropriate change of variable,

$$h(D_A, z) = (1+z)D_A, \quad \zeta(z) = \frac{1+\Omega_m z(3+3z+z^2)}{1-\Omega_m}, \tag{6.15}$$

Eq. (6.14) turns out to be the hypergeometric equation,

$$(1-\zeta)\zeta\frac{d^2 h}{d\zeta^2} + \left(\frac{1}{2} - \frac{7}{6}\zeta\right)\frac{dh}{d\zeta} + \frac{\nu(\nu+1)}{36}h = 0, \tag{6.16}$$

where the parameter ν is related to the amount of clumpiness, α, by

$$\alpha = \frac{1}{6}(3+\nu)(2-\nu); \tag{6.17}$$

the corresponding range of ν is $0 \leq \nu \leq 2$, so that with $\nu = 0\,(\alpha = 1)$ one recovers the FRW limit and $\nu = 2(\alpha = 0)$ corresponds to a totally clumped universe.

To connect the angular diameter distance D_A with the observable luminosity distance D_L, we used the Etherington relation [7],

$$D_L = D_A(1+z)^2, \tag{6.18}$$

then $D_L = D_A(1+z)^2$, D_A being solution of the hypergeometric equation. We have used the so derived ZKDR luminosity distance, with D_A from the solution to the hypergeometric equation, to contrast with observations.

6.4 Testing the ZKDR Distance vs. Observations

In practice, in the general case, Eq. (6.14) with appropriate initial conditions is solved numerically. In this work we use the solution to Eq. (6.14) in terms of hypergeometric functions, and the relation to the luminosity distance, to obtain the best fits for the matter density parameter Ω_m and the ν parameter using a chi-square (χ^2) minimization method. Moreover, the ZKDR luminosity distance at short distances (redshifts

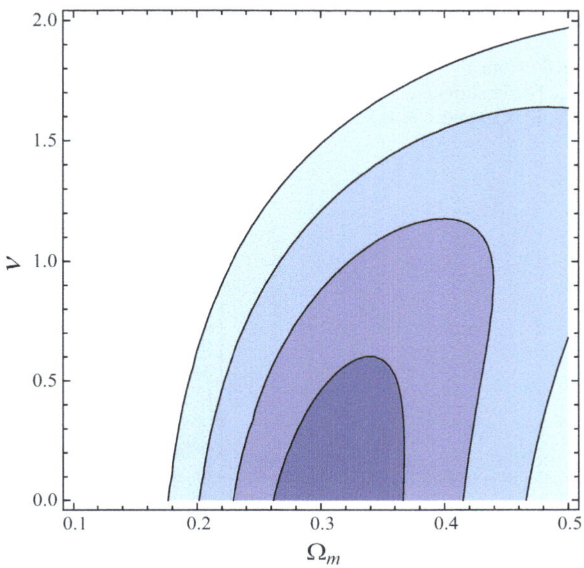

Fig. 6.1 Confidence contours in the plane (Ω_m, ν) using 26 GRBs from Schaefer [8]. The best fit is for $\Omega_m = 0.31$ and $\nu = -0.5$

in the range $0.1 < z < 1.7$) has not been shown to make a difference respect to the ΛCDM model; it is then appealing to test the model for larger distances, like the ones for GRBs, that extend the redshift range as far as 8.1.

We did the adjustment for two GRBs samples, one of 26 objects, extracted from the data presented in Schaefer [8], and the other from Y. Wang [9]. The characteristic of the employed samples is that they are model independent, in contrast with the calibration based on extrapolating the SNe Ia behaviour. For the first sample we do not get a confident fit, since the obtained ν is out of the allowed range of $0 \le \nu \le 2$. For the second sample, we get a better result that indicates that indeed there is an accumulative effect of the inhomogeneities as the light beam travels from the source to the observer, being the obtained best fits ($\Omega_m = 0.33$, $\nu = 0.84$).

The obtained confidence contours are shown in Figs. 6.1 and 6.2, constraining Ω_Λ so that $\Omega_\Lambda = 1.0 - \Omega_m$. The result for $\nu = 0$ is clearly consistent with the fraction of energy density in a flat universe attributed to dark energy in the ΛCDM model, namely, $\Omega_m = 0.3$ and $\Omega_\Lambda = 0.7$.

However, this result should be taken with caution due to the high dispersion in the GRBs data; it is well known that a robust analysis should be based on the combination of several independent cosmological data.

To complete the study we perform tests with two additional probes, namely: 557 Supernovae Ia (SNe Ia) data from the Union2 sample [10], and the measurements of Hubble parameter, based of the aging of galaxies, with a sample of 18 data reported in [11]. These two last probes are in the ranges of $0.1 < z < 1.75$, therefore the data correspond to a more recent epoch than the ones for GRBs. The adjustments are shown in Fig. 6.3 for the Union2 data and in Fig. 6.4 for the Hubble parameter.

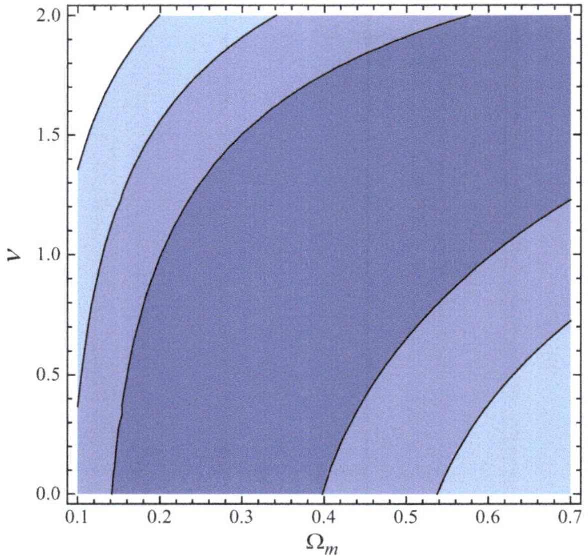

Fig. 6.2 Confidence contours in the plane (Ω_m, ν) using 59 GRBs from Y. Wang (2008) [9]. The best fits are $\Omega_m = 0.33$ and $\nu = 0.84$

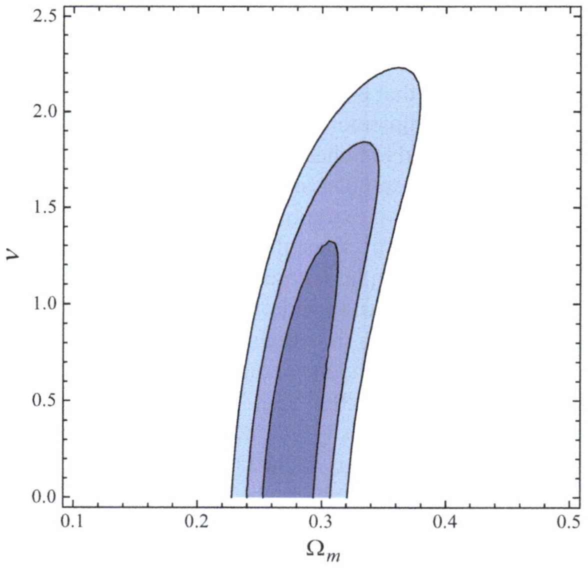

Fig. 6.3 Confidence contours in the plane (Ω_m, ν) using the 557 Union2 sample from [10]. The best fits are $\Omega_m = 0.27$ and $\nu = 0$

The best fits are $\Omega_m = 0.27$ and $\nu \approx 0$ and $\Omega_m = 0.24$ and $\nu \approx 0$, respectively. With these probes the parameter ν is constrained to the intervals $0 \leq \nu \leq 1.2$ and $0 \leq \nu \leq 0.8$, within 1σ confidence level, for SNe Ia and Hubble parameter, respectively. In fact the last two are the most confident constraints in the redshift interval $0.1 \leq z \leq 1.7$.

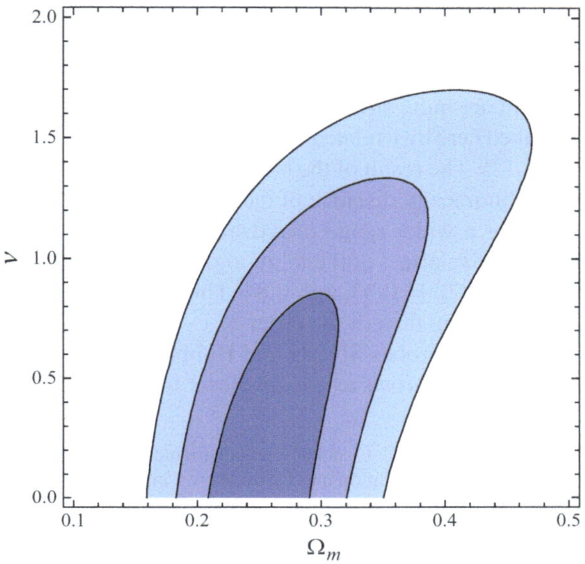

Fig. 6.4 Confidence contours in the plane (Ω_m, ν) using 18 data for the observational Hubble parameter from Moresco (2012) [11]. The best fits are $\Omega_m = 0.24$ and $\nu = 0$

From these probes we may conclude that light travel is not affected in a relevant way by the local clumpiness of the matter density. Being the range of redshift wider for the GRBs sample, $1.4 < z < 8.1$, the differences in the adjustment for ν can be attributed to an accumulative effect on the light travel, being the covered distance by the light beam one order of magnitude larger.

Our study is in agreement with previous related works. In [12], constraints on the smoothness parameter α and dark energy are found using observational $H(z)$, from samples in [13] and [14]; their analysis restricts the parameters to the intervals $1.01 < \nu < 2$ and $0.27 < \Omega_m < 0.37$ within the 2σ confidence level and the best fits are $\nu = 0$, $\Omega_m = 0.32$ in agreement with our results. A similar analysis is presented in [15], with SNe Ia and GRBs probes, using samples from the Union2 Compilation data [10] and the Hymnium GRBs [16], turning out the following constraints, $\nu \leq 1.67$ and $\Omega_m \approx 0.27$ from the SNe Ia, and from GRBs of $\Omega_m < 0.35$ while ν remained unconstrained.

6.5 Conclusions

In a flat, homogeneous and isotropic geometry (RW), the effect in the luminosity distance of local nonhomogeneous distribution of matter is measured by a phenomenological parameter ν. In order to determine Ω_m and Ω_Λ from even higher redshift observations, the distance-enhancing effect of ν is accounted for in the luminosity distance formulae proposed by Dyer and Roeder (ZKDR).

We tested the ZKDR luminosity distance with two samples of GRBs as well as with SNe Ia and Hubble parameter measurements. From this analysis it follows that variations in the angular diameter distance caused by the presence of cosmological constant are quite similar to the effects of a nonhomogeneous distribution of matter described here by ν related to the clumpiness parameter α of Dyer and Roeder through Eq. (6.17). The result of the best fits for near epochs, $0.1 < z < 1.7$, is that the effect on the luminosity distance of the local clumpiness can be neglected; however, when tested for a wider range of redshifts, $1.4 < z < 8.1$, i.e. for furthest epochs, an accumulative effect of the inhomogeneities could be the explanation of the obtained best fits of $\Omega_m = 0.33$, $\nu = 0.84$. This means that dark energy was not so important as today, being then of the order of $1 - \Omega_m = 0.67 = \Omega_\Lambda$; and combining with the result for the probes SNe Ia and Hubble parameter, it indicates that the abundance of dark energy grows as we get closer to the present time.

Acknowledgments C. C. acknowledges financial support by CINVESTAV-IPN to attend the IV Int. Meeting on Gravitation and Cosmology, Guadalajara, Jal. He also thanks a M. Sc. fellowship by CONACyT-Mexico. The authors acknowledge the referee of this work for a careful reading and useful suggestions to improve the text.

References

1. P. Astier, R. Pain, *Observational Evidence of the accelerated expansion of the universe*, (2012), arXiv:1204.5493
2. R. K. Sachs, J. Kristian, *Observations in Cosmology*, Astrophys. J. **143**, 379 (1966)
3. M. Demianski, R. de Ritis, A. A. Marino, E. Piedipalumbo, *Approximate diameter distance in a locally inhomogeneous universe with nonzero cosmological constant*, Astron. Astrophys. **411**, 33 (2003)
4. C. C. Dyer, R. C. Roeder, *Distance-redshift relations for universes with some intergalactic medium*, Astrophys. J. **174**, L115 (1972)
5. R. Kantowski, *The effects of inhomogeneities on evaluating the mass parameter and the cosmological constant*, Astrophys. J. **507**, 483 (1998)
6. R. Kantowski and R. C. Thomas, *Distance-redshift in inhomogeneous $\Omega_0 = 1$ Friedmann–Lemaitre–Robertson– Walker cosmology*, Astrophys. J. **561**, 491 (2001)
7. I. M. H. Etherington, *On the definition of distance in general relativity*, Phil. Mag. Ser. 7 **15**, 761 (1933)
8. B. E. Schaefer, *The Hubble diagram to redshift > 6 from 69 gamma-ray bursts*, Astrophys. J. **660**, 16 (2007)
9. Y. Wang, *Model-independent distance measurements for gamma-ray bursts and constraints to dark energy*, Phys. Rev. D. **78**, 123532 (2008)
10. R. Amanullah et al, *Spectra and Hubble space Telescope light curves of six type Supernovae at* $0.511 < z < 1.12$ *and the Union compilation*, Astrophys. J. **716**, 712 (2010)
11. M. Moresco, L. Verde, L. Pozzetti, R. Jimenez, A. Cimatti, *New constraints on cosmological parameters and neutrino properties using the expansion rate of the universe to* $z < 1.75$, JCAP, **07**, 053 (2012); arXiv:1201.6658
12. V. C. Busti, R. C. Santos, *Comment on "Constraining the smoothness parameter and dark energy using observational $H(z)$ data"*, Res. Astron. Astrophys. **11**, 637 (2011)
13. J. Simon, L. Verde, R. Jimenez, *Constraints on the redshift dependence of the dark energy potential*, Phys. Rev. D. **77**, 123001 (2008)

14. R. A. Daly, S. G. Djorgovski, K. A. Freeman et al., *Improved Constraints on the Acceleration History of the Universe and the Properties of the Dark Energy*, Astrophys. J. **677**, 1 (2008)
15. V. C. Busti, R. C. Santos, J. A. S. Lima, *Constraining the dark energy and smoothness parameter with SNe Ia and Gamma-Ray Bursts*, Phys. Rev. D. **85**, 103503 (2012); arXiv: 1202.0449
16. H. Wei, *Observational constraints on cosmological models with the updated long gamma-ray bursts*, JCAP. **08**, 020 (2010); arXiv1004.4951

Chapter 7
Non Singular Origin of the Universe and its Connection to the Cosmological Constant Problem (CCP)

E. I. Guendelman

Abstract We consider a non singular origin for the Universe starting from an Einstein static Universe in the framework of a theory which uses two volume elements $\sqrt{-g}d^4x$ and Φd^4x, where Φ is a metric independent density, also curvature, curvature square terms, first order formalism and for scale invariance a dilaton field ϕ are considered in the action. In the Einstein frame we also add a cosmological term that parametrizes the zero point fluctuations. The resulting effective potential for the dilaton contains two flat regions, for $\phi \to \infty$ relevant for the non singular origin of the Universe and $\phi \to -\infty$, describing our present Universe. Surprisingly, avoidance of singularities and stability as $\phi \to \infty$ imply a positive but small vacuum energy as $\phi \to -\infty$. Zero vacuum energy density for the present universe is the "threshold" for universe creation.

7.1 Introduction

The "Cosmological Constant Problem" [1–3] (CCP), is a consequence of the uncontrolled UV behavior of the zero point fluctuations in Quantum Field Theory (QFT), which leads to an equally uncontrolled vacuum energy density or cosmological constant term (CCT). This CCT is undetermined in QFT, but it is naturally very large, unless a delicate balance of huge quantities, for some unknown reason, conspires to give a very small final result. Here we will explore a candidate mechanism where the CCT is controlled, in a the context of a very specific framework, by the requirement of a non singular origin for the universe.

We will adopt the very attractive "Emergent Universe" scenario, where conclusions concerning singularity theorems can be avoided [4–11] by violating the geometrical assumptions of these theorems. In this scenario [4, 5] we start at very early times ($t \to -\infty$) with a closed static Universe (Einstein Universe).

E. I. Guendelman (✉)
Physics Department, Ben Gurion University of the Negev, Beer Sheva, 84105, Israel
e-mail: guendel@bgu.ac.il

C. Moreno González et al. (eds.), *Accelerated Cosmic Expansion*,
Astrophysics and Space Science Proceedings 38,
DOI 10.1007/978-3-319-02063-1_7,

In [4] even models based on standard General Relativity, ordinary matter and minimally coupled scalar fields were considered and can provide indeed a non singular (geodesically complete) inflationary universe, with a past eternal Einstein static Universe that eventually evolves into an inflationary Universe.

Those most simple models suffer however from instabilities, associated with the instability of the Einstein static universe. The instability is possible to cure by going away from GR, considering non perturbative corrections to the Einstein's field equations in the context of the loop quantum gravity [6], a brane world cosmology [7], considering the Starobinsky model for radiative corrections (which cannot be derived from an effective action) [8] or exotic matter[9]. In addition to this, the consideration of a Jordan Brans Dicke model also can provide a stable initial state for the emerging universe scenario [10, 11].

Here we discuss a different theoretical framework, presented in details in ref. [12], where such emerging universe scenario is realized in a natural way, where instabilities are avoided and an inflationary phase with a graceful exit can be achieved. The model we will use was studied first in [13] (in ref. [12] a few typos in [13] have been corrected and also the discussion of some notions discussed there as well has been improved), however, we differ with [13] in our choice of the state with (here and in ref. [12] with a lower vacuum energy density) that best represents the present state of the universe. This is crucial, since as it should be obvious, the discussion of the CCP depends crucially on what vacuum we take. We will express the stability and existence conditions for the non singular initial universe in terms of the energy of the vacuum of our candidate for the present Universe.

7.2 The Model

We work in the context of a theory built along the lines of the two measures theory (TMT) [14–17] which deals with actions of the form,

$$S = \int L_1 \sqrt{-g} d^4x + \int L_2 \Phi d^4x \tag{7.1}$$

where Φ is an alternative "measure of integration", a density independent of the metric, for example in terms of four scalars φ_a ($a = 1, 2, 3, 4$), it can be obtained as follows:

$$\Phi = \varepsilon^{\mu\nu\alpha\beta} \varepsilon_{abcd} \partial_\mu \varphi_a \partial_\nu \varphi_b \partial_\alpha \varphi_c \partial_\beta \varphi_d \tag{7.2}$$

and more specifically work in the context of the globally scale invariant realization of such theories [15, 16], which require the introduction of a dilaton field ϕ. We look at the generalization of these models [16] where an "R^2 term" is present,

$$L_1 = U(\phi) + \varepsilon R(\Gamma, g)^2 \tag{7.3}$$

$$L_2 = \frac{-1}{\kappa} R(\Gamma, g) + \frac{1}{2} g^{\mu\nu} \partial_\mu \phi \partial_\nu \phi - V(\phi) \tag{7.4}$$

$$R(\Gamma, g) = g^{\mu\nu} R_{\mu\nu}(\Gamma), R_{\mu\nu}(\Gamma) = R^{\lambda}_{\mu\nu\lambda} \tag{7.5}$$

$$R^{\lambda}_{\mu\nu\sigma}(\Gamma) = \Gamma^{\lambda}_{\mu\nu,\sigma} - \Gamma^{\lambda}_{\mu\sigma,\nu} + \Gamma^{\lambda}_{\alpha\sigma}\Gamma^{\alpha}_{\mu\nu} - \Gamma^{\lambda}_{\alpha\nu}\Gamma^{\alpha}_{\mu\sigma}. \tag{7.6}$$

global scale invariance is satisfied if [16, 15] (f_1, f_2, α being constants),

$$V(\phi) = f_1 e^{\alpha\phi}, U(\phi) = f_2 e^{2\alpha\phi} \tag{7.7}$$

In the variational principle $\Gamma^{\lambda}_{\mu\nu}$, $g_{\mu\nu}$, the measure fields scalars φ_a and the "matter"—scalar field ϕ are all to be treated as independent variables although the variational principle may result in equations that allow us to solve some of these variables in terms of others, that is, the first order formalism is employed, where any relation between the connection coefficients and the metric is obtained from the variational principle, not postulated a priori. A particularly interesting equation is the one that arises from the φ_a fields, this yields $L_2 = M$, where M is a constant that spontaneously breaks scale invariance. the Einstein frame, which is a redefinition of the metric by a conformal factor, is defined as

$$\overline{g}_{\mu\nu} = (\chi - 2\kappa\varepsilon R) g_{\mu\nu} \tag{7.8}$$

where χ is the ratio between the two measures, $\chi = \frac{\Phi}{\sqrt{-g}}$, determined from the consistency of the equations to be $\chi = \frac{2U(\phi)}{M+V(\phi)}$. The relevant fact is that the connection coefficient equals the Christoffel symbol of this new metric (for the original metric this "Riemannian" relation does not hold).

One could question the use of the Einstein frame metric $\overline{g}_{\mu\nu}$ in contrast to the original metric $g_{\mu\nu}$. In this respect, it is interesting to see the role of both the original metric and that of the Einstein frame metric in a canonical approach to the first order formalism. Here we see that the original metric does not have a canonically conjugated momentum (this turns out to be zero), in contrast, the canonically conjugated momentum to the connection turns out to be a function exclusively of $\overline{g}_{\mu\nu}$, this Einstein metric is therefore a genuine dynamical canonical variable, as opposed to the original metric. There is also a lagrangian formulation of the theory which uses $\overline{g}_{\mu\nu}$, as we will see in the next section, what we can call the action in the Einstein frame. In this frame we can quantize the theory for example and consider contributions without reference to the original frame, thus possibly considering breaking the TMT structure of the theory through quantum effects, but such breaking will be done "softly" through the introduction of a cosmological term only. Surprisingly, the remaining structure of the theory, reminiscent from the original TMT structure will be enough to control the strength of this additional cosmological term once we demand that the universe originated from a non singular and stable emergent state.

There is a "k-essence" type effective action, where one can use this Einstein frame metric.

As it is standard in treatments of theories with non linear kinetic terms or k-essence models [18–21], it is determined by a pressure functional, ($X = \frac{1}{2}\overline{g}^{\mu\nu}\partial_\mu\phi\partial_\nu\phi$).

The effective energy-momentum tensor can be represented in a form like that of a perfect fluid

$$T^{eff}_{\mu\nu} = (\rho + p)u_\mu u_\nu - p\tilde{g}_{\mu\nu}, \qquad \text{where} \qquad u_\mu = \frac{\phi_{,\mu}}{(2X)^{1/2}} \tag{7.9}$$

here $X \equiv \frac{1}{2}\tilde{g}^{\alpha\beta}\phi_{,\alpha}\phi_{,\beta}$. This defines a pressure functional and an energy density functional. The system of equations obtained after solving for χ, working in the Einstein frame with the metric $\tilde{g}_{\mu\nu}$ can be obtained from a "k-essence" type effective action, as it is standard in treatments of theories with non linear kinetic terms or k-essence models [18–21]. The action from which the classical equations follow is,

$$S_{eff} = \int \sqrt{-\tilde{g}}d^4x \left[-\frac{1}{\kappa}\tilde{R}(\tilde{g}) + p(\phi, R)\right] \tag{7.10}$$

$$p = \frac{\chi}{\chi - 2\kappa\varepsilon R}X - V_{eff} \tag{7.11}$$

$$V_{eff} = \frac{\varepsilon R^2 + U}{(\chi - 2\kappa\varepsilon R)^2} \tag{7.12}$$

$\overline{R}$ is the Riemanniann curvature scalar built out of the bar metric, R on the other hand is the non Riemannian curvature scalar defined in terms of the connection and the original metric, which turns out to be given by $R = \frac{-\kappa(V+M)+\frac{\kappa}{2}\overline{g}^{\mu\nu}\partial_\mu\phi\partial_\nu\phi\chi}{1+\kappa^2\varepsilon\overline{g}^{\mu\nu}\partial_\mu\phi\partial_\nu\phi}$. This R can be inserted in the action or alternatively, R in the action can be treated as an independent degree of freedom, then its variation gives the required value as one can check (which can then be reinserted in the action. Introducing this R into the expression (7.12) and considering a constant field ϕ we find that V_{eff} has two flat regions. The existence of two flat regions for the potential is shown to be consequence of the s.s.b. of the scale symmetry (that is of considering $M \neq 0$).

We will focus only on a possible cosmological term in the Einstein frame added (due to zero point fluctuations) to (7.10), which leads then to the new action

$$S_{eff,\Lambda} = \int \sqrt{-\overline{g}}d^4x \left[-\frac{1}{\kappa}\overline{R}(\overline{g}) + p(\phi, R) - \Lambda\right] \tag{7.13}$$

This addition to the effective action leaves the equations of motion of the scalar field unaffected, but the gravitational equations aquire a cosmological constant. Adding the Λ term can be regarded as a redefinition of $V_{eff}(\phi, R)$

$$V_{eff}(\phi, R) \rightarrow V_{eff}(\phi, R) + \Lambda \tag{7.14}$$

7.3 Emergent Solutions and Their Stability

We start considering the cosmological solutions of the form

$$ds^2 = dt^2 - a(t)^2\left(\frac{dr^2}{1-r^2} + r^2(d\theta^2 + sin^2\theta d\phi^2)\right), \phi = \phi(t) \tag{7.15}$$

in this case, we obtain for the energy density and the pressure, the following expressions. We will consider a scenario where the scalar field ϕ is moving in the extreme right region $\phi \to \infty$, in this case the expressions for the energy density ρ and pressure p are given by,

$$\rho = \frac{A}{2}\dot{\phi}^2 + 3B\dot{\phi}^4 + C \tag{7.16}$$

and

$$p = \frac{A}{2}\dot{\phi}^2 + B\dot{\phi}^4 - C \tag{7.17}$$

It is interesting to notice that all terms proportional to $\dot{\phi}^4$ behave like "radiation", since $p_{\dot{\phi}^4} = \frac{\rho_{\dot{\phi}^4}}{3}$ is satisfied. here the constants A, B and C are given by,

$$A = \frac{f_2}{f_2 + \kappa^2\varepsilon f_1^2}, \tag{7.18}$$

$$B = \frac{\varepsilon\kappa^2}{4(1 + \kappa^2\varepsilon f_1^2/f_2)} = \frac{\varepsilon\kappa^2}{4}A, \tag{7.19}$$

$$C = \frac{f_1^2}{4f_2}A + \Lambda. \tag{7.20}$$

It will be convenient to "decompose" the constant Λ into two pieces,

$$\Lambda = -\frac{1}{4\kappa^2\varepsilon} + \Delta\lambda \tag{7.21}$$

since as $\phi \to -\infty$, $V_{eff} \to \Delta\lambda$. Therefore $\Delta\lambda$ has the interesting interpretation of the vacuum energy density in the $\phi \to -\infty$ vacuum.

The equation that determines such static universe $a(t) = a_0 = constant, \dot{a} = 0$, $\ddot{a} = 0$ gives rise to a restriction for $\dot{\phi}_0$ that have to satisfy the following equation in order to guarantee that the universe be static, because $\ddot{a} = 0$ is proportional to $\rho + 3p$, we must require that $\rho + 3p = 0$, which leads to

$$3B\dot{\phi}_0^4 + A\dot{\phi}_0^2 - C = 0, \tag{7.22}$$

This equation leads to two roots, the first being

$$\dot{\phi}_1^2 = \frac{\sqrt{A^2 + 12BC} - A}{6B}. \tag{7.23}$$

The second root is:

$$\dot{\phi}_2^2 = \frac{-\sqrt{A^2 + 12BC} - A}{6B}. \tag{7.24}$$

It is also interesting to see that if the discriminant is positive, the first solution has automatically positive energy density, if we only consider cases where $C > 0$, which is required if we want the emerging solution to be able to turn into an inflationary solution eventually. One can see that the condition $\rho > 0$ for the first solution reduces to the inequality $w > (1 - \sqrt{1-w})/2$, where $w = -12BC/A^2 > 0$, since we must have $A > 0$, otherwise we get a negative kinetic term during the inflationary period, and as we will see in the next section, we must have that $B < 0$ from the stability of the solution, and as long as $w < 1$, it is always true that this inequality is satisfied.

We will now consider the perturbation equations. Considering small deviations of $\dot{\phi}$ the from the static emergent solution value $\dot{\phi}_0$ and also considering the perturbations of the scale factor a, we obtain, from Eq. (7.16)

$$\delta\rho = A\dot{\phi}_0\delta\dot{\phi} + 12B\dot{\phi}_0^3\delta\dot{\phi} \tag{7.25}$$

at the same time $\delta\rho$ can be obtained from the perturbation of the Friedmann equation

$$3(\frac{1}{a^2} + H^2) = \kappa\rho \tag{7.26}$$

and since we are perturbing a solution which is static, i.e., has $H = 0$, we obtain then

$$-\frac{6}{a_0^3}\delta a = \kappa\delta\rho \tag{7.27}$$

we also have the second order Friedmann equation

$$\frac{1 + \dot{a}^2 + 2a\ddot{a}}{a^2} = -\kappa p \tag{7.28}$$

For the static emerging solution, we have $p_0 = -\rho_0/3$, $a = a_0$, so

$$\frac{2}{a_0^2} = -2\kappa p_0 = \frac{2}{3}\kappa\rho_0 = \Omega_0\kappa\rho_0 \tag{7.29}$$

where we have chosen to express our result in terms of Ω_0, defined by $p_0 = (\Omega_0 - 1)$ ρ_0, which for the emerging solution has the value $\Omega_0 = \frac{2}{3}$. Using this in 7.27, we obtain

$$\delta\rho = -\frac{3\Omega_0\rho_0}{a_0}\delta a \tag{7.30}$$

and equating the values of $\delta\rho$ as given by (7.25) and (7.30) we obtain a linear relation between $\delta\dot{\phi}$ and δa, which is,

$$\delta\dot{\phi} = D_0\delta a \tag{7.31}$$

where

$$D_0 = -\frac{3\Omega_0\rho_0}{a_0\dot{\phi}_0(A + 12B\dot{\phi}_0^2)} \tag{7.32}$$

we now consider the perturbation of the Eq. (7.28). In the right hand side of this equation we consider that $p = (\Omega - 1)\rho$, with

$$\Omega = 2\big(1 - \frac{U_{eff}}{\rho}\big), \tag{7.33}$$

where,

$$U_{eff} = C + B\,\dot{\phi}^4 \tag{7.34}$$

and therefore, the perturbation of the Eq. (7.28) leads to,

$$-\frac{2\delta a}{a_0^3} + 2\frac{\delta\ddot{a}}{a_0} = -\kappa\delta p = -\kappa\delta((\Omega - 1)\rho) \tag{7.35}$$

to evaluate this, we use (7.33, 7.34) and the expressions that relate the variations in a and $\dot{\phi}$ (7.31). Defining the "small" variable β as

$$a(t) = a_0(1 + \beta) \tag{7.36}$$

we obtain,

$$2\ddot{\beta}(t) + W_0^2\beta(t) = 0, \tag{7.37}$$

where,

$$W_0^2 = \Omega_0\,\rho_0\left[\frac{24\,B\,\dot{\phi}_0^2}{A + 12\,\dot{\phi}_0^2\,B} - 6\frac{(C + B\,\dot{\phi}_0^4)}{\rho_0} - 3\kappa\,\Omega_0 + 2\kappa\right], \tag{7.38}$$

notice that the sum of the last two terms in the expression for W_0^2, that is $-3\kappa\,\Omega_0 + 2\kappa$ vanish since $\Omega_0 = \frac{2}{3}$, for the same reason, we have that $6\frac{(C+B\,\dot{\phi}_0^4)}{\rho_0} = 4$, which brings us to the simplified expression

$$W_0^2 = \Omega_0\,\rho_0\left[\frac{24\,B\,\dot{\phi}_0^2}{A + 12\,\dot{\phi}_0^2\,B} - 4\right], \tag{7.39}$$

For the stability of the static solution, we need that $W_0^2 > 0$, where $\dot{\phi}_0^2$ is defined either by Eq. (7.23) ($\dot{\phi}_0^2 = \phi_1^2$) or by Eq. (7.24) ($\dot{\phi}_0^2 = \phi_2^2$). If we take Eq. (7.24) ($\dot{\phi}_0^2 = \phi_2^2$) and use this in the above expression for W_0^2, we obtain,

$$W_0^2 = \Omega_0\,\rho_0\left[\frac{4\sqrt{A^2 + 12BC}}{-2\sqrt{A^2 + 12BC} - A}\right], \tag{7.40}$$

to avoid negative kinetic terms during the slow roll phase that takes place following the emergent phase, we must consider $A > 0$, so, we see that the second solution is unstable and will not be considered further.

Now in the case of the first solution, Eq. (7.23) ($\dot{\phi}_0^2 = \phi_1^2$), then W_0^2 becomes

$$W_0^2 = \Omega_0 \, \rho_0 \left[\frac{-4\sqrt{A^2 + 12BC}}{2\sqrt{A^2 + 12BC} - A} \right], \tag{7.41}$$

so the condition of stability becomes $2\sqrt{A^2 + 12BC} - A < 0$, or $2\sqrt{A^2 + 12BC} < A$,squaring both sides and since $A > 0$, we get $12BC/A^2 < -3/4$, which means $B < 0$, and therefore $\varepsilon < 0$, multiplying by -1, we obtain, $12(-B)C/A^2 > 3/4$, replacing the values of A, B, C, given by 7.18 we obtain the condition

$$\Delta\lambda > 0, \tag{7.42}$$

Now there is the condition that the discriminant be positive $A^2 + 12BC > 0$

$$\Delta\lambda < \frac{1}{12(-\varepsilon)\kappa^2} \left[\frac{f_2}{f_2 + \kappa^2 \varepsilon f_1^2} \right], \tag{7.43}$$

since $A = \left[\frac{f_2}{f_2 + \kappa^2 \varepsilon f_1^2} \right] > 0$, $B < 0$, meaning that $\varepsilon < 0$, we see that we obtain a positive upper bound for the energy density of the vacuum as $\phi \to -\infty$, which must be positive, but not very big.

These are very satisfactory results, since it means that the existence and stability of the emerging universe prevents the vacuum energy in the present universe from being very large, but requires it to be positive. The transition from the emergent universe to the ground state goes through an intermediate inflationary phase, therefore reproducing the basic standard cosmological model as well. So, it turns out that the creation of the universe can be considered as a "threshold event" for zero present vacuum energy density, which naturally gives a positive but small vacuum energy density for the present universe.

Acknowledgments I would like to thank the organizers of this very interesting conference, to Salvatore Capozziello for very useful discussions on the special role of the Einstein frame.

References

1. S. Weinberg, Rev. Mod. Phys. **61**, 1 (1989)
2. Y. Jack Ng. Int. J. Mod. Phys. **D1**, 145 (1992)
3. S. Weinberg, astro-ph/0005265
4. G.F.R. Ellis, R. Maartens, Class. Quantum Grav. **21**, 223 (2004)
5. G.F.R. Ellis, J. Murugan, C.G. Tsagas, Class. Quantum Grav. **21**, 233 (2004)
6. D.J. Mulryne, R. Tavakol, J.E. Lidsey, G.F.R. Ellis, Phys. Rev. **D71**, 123512 (2005)
7. A. Banerjee, T. Bandyopadhyay, S. Chakraborty, Grav.Cosmol. **13**, 290 (2007)

8. S. Mukherjee, B.C. Paul, S.D. Maharaj, A. Beesham, arXiv:qr-qc/0505103
9. S. Mukherjee, B.C. Paul, N.K. Dadhich, S.D. Maharaj, A. Beesham, Class. Quantum Grav. **23**, 6927 (2006)
10. S. del Campo, R. Herrera, P. Labrana, JCAP **0907**, 006 (2009)
11. S. del Campo, R. Herrera, P. Labrana, JCAP **0711**, 030 (2007)
12. E.I. Guendelman, Int. J. Mod. Phys. **A26** 2951 (2011), e-Print: arXiv:1103.1427 [gr-qc]
13. S. del Campo, E.I. Guendelman, R. Herrera, P. Labrana, JCAP **1006**, 026 (2010)
14. Basic idea is developed in E.I. Guendelman, A.B. Kaganovich, Phys. Rev. **D60**, 065004 (1999)
15. For a recent review and further references see E.I. Guendelman, A.B. Kaganovich, Plenary talk given at the Workshop on Geometry, Topology, QFT and Cosmology, Paris, France, 28-30 May 2008. e-Print: arXiv:0811.0793 [gr-qc]
16. E.I. Guendelman, Mod. Phys. Lett. **A14**, 1043 (1999), e-Print: gr-qc/9901017
17. E.I. Guendelman, O. Katz, Class. Quantum Grav. **20**, 1715 (2003), e-Print: gr-qc/0211095
18. T. Chiba, T.Okabe, M. Yamaguchi, Phys.Rev. **D62**, 023511 (2000)
19. C. Armendariz-Picon, V. Mukhanov, P.J. Steinhardt, Phys.Rev.Lett. **85**, 4438 2000
20. C. Armendariz-Picon, V. Mukhanov, P.J. Steinhardt, Phys.Rev. **D63**, 103510 (2001)
21. T. Chiba, Phys.Rev. **D66** (063514) (2002)

Chapter 8
Tunneling and the Emergent Universe Scheme

Pedro Labraña

Abstract We present an alternative scheme for an Emergent Universe scenario, developed previously in Phys. Rev. D **86**, 083524 (2012), where the universe is initially in a static state supported by a scalar field located in a false vacuum. The universe begins to evolve when, by quantum tunneling, the scalar field decays into a state of true vacuum.

8.1 Introduction

Singularity theorems have been devised that apply in the inflationary context, showing that the universe necessarily had a beginning [2–6]. However, recently, models that escape this conclusion has been studied in Refs. [7–14]. These models do not satisfy the geometrical assumptions of these theorems. Specifically, the theorems assume that either (1) the universe has open space sections, implying $k = 0$ or -1, or (2) the Hubble expansion rate H is bounded away from zero in the past, $H > 0$. Such models, called Emergent Universe, are appealing since they provide specific examples of non-singular (geodesically complete) inflationary universes.

Normally in the Emergent Universe scenario, the universe is positively curved and initially it is in a past eternal classical Einstein static state which eventually evolves into a subsequent inflationary phase, see [7–14].

For example, in the original scheme [7, 8], it is assumed that the universe is dominated by a scalar field (inflaton) ϕ with a scalar potential $V(\phi)$ that approach a constant V_0 as $\phi \rightarrow -\infty$ and monotonically rise once the scalar field exceeds a certain value ϕ_0, see Fig. 8.1.

During the past-eternal static regime it is assumed that the scalar field is rolling on the asymptotically flat part of the scalar potential with a constant velocity, providing the conditions for a static universe. But once the scalar field exceeds some value, the scalar potential slowly droops from its original value. The overall effect of this is to distort the equilibrium behavior breaking the static solution. If the potential has a suitable form in this region, slow-roll inflation will occur, thereby providing a 'graceful entrance' to early universe inflation.

P. Labraña (✉)
Departamento de Física, Universidad del Bío-Bío,
Avenida Collao 1202, Casilla 5-C, Concepción, Chile
e-mail: plabrana@ubiobio.cl

C. Moreno González et al. (eds.), *Accelerated Cosmic Expansion*,
Astrophysics and Space Science Proceedings 38,
DOI 10.1007/978-3-319-02063-1_8,

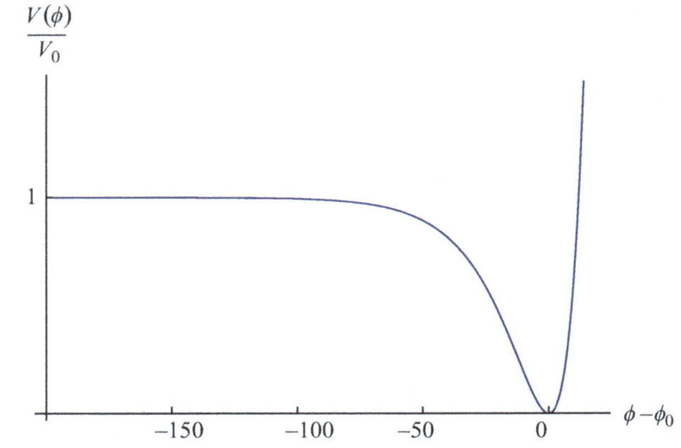

Fig. 8.1 Schematic representation of a potential for a standard Emergent Universe scenario

Notice that, as was shown by Eddington [15], the Einstein static state is unstable to homogeneous perturbations. This situation has implication for the Emergent Universe scenario, see discussion in Sect. 8.5.

This scheme for a Emergent Universe have been used not only on models based on General Relativity [7, 8], but also on models where non-perturbative quantum corrections of the Einstein field equations are considered [9, 13, 14], in the context of a scalar tensor theory of gravity [16, 17] and recently in the framework of the so-called two measures field theories [18–21].

Another possibility for the Emergent Universe scenario is to consider models in which the scale factor asymptotically tends to a constant in the past [10, 11, 22–27].

We can note that both schemes for a Emergent Universe are not truly static during the static regime. For instance, in the first scheme during the static regime the scalar field is rolling on the flat part of its potential. On the other hand, for the second scheme the scale factor is only asymptotically static.

However, recently, it has been proposed an alternative scheme for an Emergent Universe scenario, where the universe is initially in a truly static state [1]. This state is supported by a scalar field which is located in a false vacuum ($\phi = \phi_F$), see Fig. 8.2. The universe begins to evolve when, by quantum tunneling, the scalar field decays into a state of true vacuum. Then, a small bubble of a new phase of field value ϕ_W can form, and expand as it converts volume from high to low vacuum energy and feeds the liberated energy into the kinetic energy of the bubble wall [33, 34]. Inside the bubble, space-like surfaces of constant ϕ are homogeneous surfaces of constant negative curvature. One way of describing this situation is to say that the interior of the bubble always contains an open Friedmann–Robertson–Walker universe [34]. If the potential has a suitable form, inflation and reheating may occur in the interior of the bubble as the field rolls from ϕ_W to the true minimum at ϕ_T, in a similar way to what happens in models of Open Inflationary Universes, see for example [28–32].

In Ref. [1] we considered a simplified version of this scheme, where we focused on studied the process of creation and evolution of a bubble of true vacuum in the

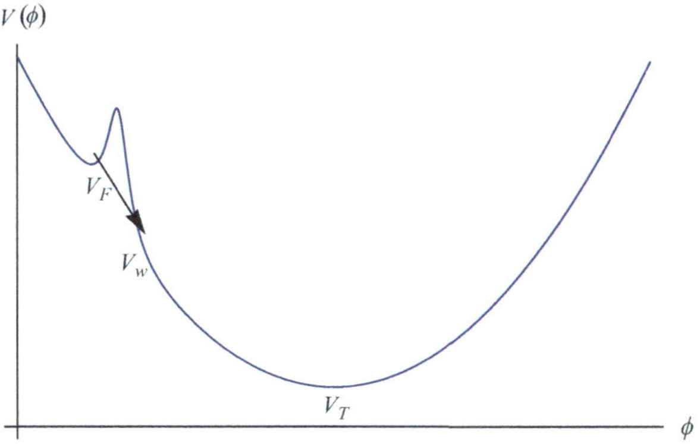

Fig. 8.2 A double-well inflationary potential $V(\phi)$. In the graph, some relevant values are indicated. They are the false vacuum $V_F = V(\phi_F)$ from which the tunneling begins, $V_W = V(\phi_W)$ where the tunneling stops and where the inflationary era begins, while $V_T = V(\phi_T)$ denote the true vacuum energy

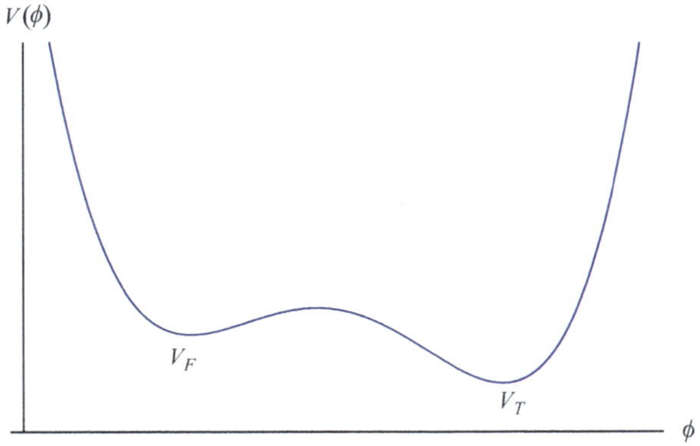

Fig. 8.3 Potential with a false and true vacuum

background of an Einstein Static (ES) universe. In particular, we considered an inflaton potential similar to Fig. 8.3 and studied the process of tunneling of the scalar field from the false vacuum ϕ_F to the true vacuum ϕ_T and the consequent creation and evolution of a bubble of true vacuum in the background of an ES universe. Here we review the principal results of Ref. [1].

8.2 Static Universe Background

Based on the standard Emergent Universe (EU) scenario, we consider that the universe is positively curved and it is initially in a past eternal classical Einstein static state. The matter of the universe is modeled by a standard perfect fluid $P = (\gamma - 1)\rho$ and a scalar field (inflaton) with energy density $\rho_\phi = \frac{1}{2}(\partial_t\phi)^2 + V(\phi)$ and pressure $P_\phi = \frac{1}{2}(\partial_t\phi)^2 - V(\phi)$. The scalar field potential $V(\phi)$ is depicted in Fig. 8.3. The global minimum of $V(\phi)$ is tiny and positive, at a field value ϕ_T, but there is also a local false minimum at $\phi = \phi_F$.

The metric for the static state is given by the closed Friedmann–Robertson–Walker metric:

$$ds^2 = dt^2 - a(t)^2 \left[\frac{dr^2}{1 - \frac{r^2}{R^2}} + r^2 (d\theta^2 + \sin^2\theta \, d\phi^2) \right], \tag{8.1}$$

where $a(t)$ is the scale factor, t represents the cosmic time and the constant $R > 0$.

We have explicitly written R in the metric in order to make more clear the effects of the curvature on the bubble process (probability of creation and propagation of the bubble).

Given that there are no interactions between the standard fluid and the scalar field, they separately obey energy conservation and Klein–Gordon equations,

$$\partial_t \rho + 3\gamma H \rho = 0, \tag{8.2}$$

$$\partial_t^2 \phi + 3H \partial_t \phi = -\frac{\partial V(\phi)}{\partial \phi}, \tag{8.3}$$

where $H = \partial_t a / a$.

The Friedmann and the Raychaudhuri field equations become,

$$H^2 = \frac{8\pi G}{3} \left(\rho + \frac{1}{2} (\partial_t \phi)^2 + V(\phi) \right) - \frac{1}{R^2 a^2}, \tag{8.4}$$

$$\partial_t^2 a = -\frac{8\pi G}{3} a \left[\left(\frac{3}{2}\gamma - 1 \right) \rho + \dot{\phi}^2 - V(\phi) \right]. \tag{8.5}$$

The static universe is characterized by the conditions $a = a_0 = const.$, $\partial_t a_0 = \partial_t^2 a_0 = 0$ and $\phi = \phi_F = Cte.$, $V(\phi_F) = V_F$ corresponding to the false vacuum.

From Eqs. (8.2) to (8.5), the static solution for a universe dominated by a scalar field placed in a false vacuum and a standard perfect fluid, are obtained if the following conditions are met

$$\rho_0 = \frac{1}{4\pi G} \frac{1}{\gamma R^2 a_0^2}, \tag{8.6}$$

$$V_F = \left(\frac{3}{2}\gamma - 1 \right) \rho_0, \tag{8.7}$$

where ρ_0 is energy density of the perfect fluid present in the static universe. Note that $\gamma > 2/3$ in order to have a positive scalar potential.

By integrating Eq. (8.2) we obtain

$$\rho = \frac{A}{a^{3\gamma}}, \tag{8.8}$$

where A is an integration constant. By using this result, we can rewrite the conditions for a static universe as follow

$$A = \frac{1}{4\pi G}\,\frac{a_0^{3\gamma-2}}{\gamma R^2}, \tag{8.9}$$

$$V_F = \left(\frac{3}{2}\gamma - 1\right)\frac{1}{4\pi G}\,\frac{1}{\gamma R^2 a_0^2}. \tag{8.10}$$

In a purely classical field theory if the universe is static and supported by the scalar field located at the false vacuum V_F, then the universe remains static forever. Quantum mechanics makes things more interesting because the field can tunnel through the barrier and by this process create a small bubble where the field value is ϕ_T. Depending of the background where the bubble materializes, the bubble could expanded or collapsed [37, 41].

8.3 Bubble Nucleation

In this section we study the tunneling process of the scalar field from the false vacuum to the true vacuum and the consequent creation of a bubble of true vacuum in the background of Einstein static universe. Given that in our case the geometry of the background correspond to a Einstein static universe and not a de Sitter space, we proceed following the scheme developed in [35, 37], instead of the usual semiclassical calculation of the nucleation rate based on instanton methods [34]. In particular, we will consider the nucleation of a spherical bubble of true vacuum V_T within the false vacuum V_F. We will assume that the layer which separates the two phases (the wall) is of negligible thickness compared to the size of the bubble (the usual thin-wall approximation). The energy budget of the bubble consists of latent heat (the difference between the energy densities of the two phases) and surface tension.

In order to eliminate the problem of predicting the reaction of the geometry to an essentially a-causal quantum jump, we neglect during this computation the gravitational back-reaction of the bubble onto the space-time geometry.

The gravitational back-reaction of the bubble will be consider in the next chapter when we study the evolution of the bubble after its materialization.

In our case the shell trajectory follows from the action (see [35, 36])

$$S = \int dy\left\{2\pi\,\varepsilon\,\bar{a}_0^4\left[\chi - \cos(\chi)\sin(\chi)\right] - 4\pi\,\sigma\,\bar{a}_0^3\,\sin^2(\chi)\sqrt{1-\chi'^2}\right\}. \tag{8.11}$$

where we have denoted the coordinate radius of the shell as χ, and we have written the static ($a = a_0 = Cte.$) version of the metric Eq. (8.1) as

$$ds^2 = \bar{a}_0^2\left(dy^2 - d\chi^2 - \sin^2(\chi)d\Omega^2\right), \tag{8.12}$$

with $\frac{r}{R} = \sin(\chi)$, $\bar{a}_0 = Ra_0$, $dt = \bar{a}_0\,dy$ and prime means derivatives respect to y.

In the action (8.11), ε and σ denote, respectively, the latent heat and the surface energy density (surface tension) of the shell.

The action (8.11) describes the classical trajectory of the shell after the tunneling. This trajectory emanates from a classical turning point, where the canonical momentum

$$P = \frac{\partial S}{\partial \chi'} = 4\pi\,\sigma\,\bar{a}_0^3\,\chi'\,\frac{\sin^2(\chi)}{\sqrt{1-\chi'^2}}, \tag{8.13}$$

vanishes [35]. In order to consider tunneling, we evolve this solution back to the turning point, and then try to shrink the bubble to zero size along a complex y contour, see [35, 37]. For each solution, the semiclassical tunneling rate is determined by the imaginary part of its action, see [35]:

$$\Gamma \approx e^{-2Im[S]}. \tag{8.14}$$

From the action (8.11) we found the equation of motion

$$\frac{\sin^2(\chi)}{\sqrt{1-\chi'^2}} = \frac{\varepsilon\,\bar{a}_0}{2\sigma}\big[\chi - \cos(\chi)\sin(\chi)\big]. \tag{8.15}$$

The action (8.11) can be put in a useful form by using Eq. (8.15), and changing variables to χ:

$$S = \int d\chi\,\frac{4\pi}{3}\,\varepsilon\,a_0^4\,\sin^2(\chi)\sqrt{\left(\frac{3[\chi - \cos(\chi)\sin(\chi)]}{2\sin^2(\chi)}\right)^2 - \bar{r}_0^2}, \tag{8.16}$$

where $\bar{r}_0 = \frac{r_0}{R}$ and $r_0 = \frac{3\sigma}{\varepsilon\,a_0}$ is the radio of nucleation of the bubble when the space is flat ($R \to \infty$) and static (i.e. when the space is Minkowsky).

The nucleation radius $\bar{\chi}$ (i.e. the coordinate radius of the bubble at the classical turning point), is a solution to the condition $P = 0$. Then from Eq. (8.13) we obtain

$$\frac{\bar{\chi} - \cos(\bar{\chi})\sin(\bar{\chi})}{\sin^2(\bar{\chi})} = \frac{2\sigma}{\varepsilon\,\bar{a}_0}. \tag{8.17}$$

The action (8.11) has an imaginary part coming from the part of the trajectory $0 < \chi < \bar{\chi}$, when the bubble is tunneling:

$$Im[S] = \frac{4\pi}{3}\,\varepsilon\,a_0^4 \int_0^{\bar{\chi}} d\chi\,\sin^2(\chi)\sqrt{\bar{r}_0^2 - \left(\frac{3[\chi - \cos(\chi)\sin(\chi)]}{2\sin^2(\chi)}\right)^2}, \tag{8.18}$$

Expanding (8.18) at first nonzero contribution in $\beta = (r_0/R)^2$ we find

$$Im[S] = \frac{27\,\sigma^4\,\pi}{4\,\varepsilon^3}\left[1 - \frac{1}{2}\beta^2\right] \tag{8.19}$$

This result is in agreement with the expansion obtained in [38]. Then, the nucleation rate is

$$\Gamma \approx e^{-2Im\,S} \approx \exp\left[-\frac{27\sigma^4\pi}{2\varepsilon^3}\left(1 - \frac{9\sigma^2}{2\varepsilon^3\,a_0^2 R^2}\right)\right]. \tag{8.20}$$

We can note that the probability of the bubble nucleation is enhanced by the effect of the curvature of the closed static universe background.

8.4 Evolution of the Bubble

In this section we study the evolution of the bubble after the process of tunneling. During this study we are going to consider the gravitational back-reaction of the bubble. We follow the approach used in [41] where it is assumed that the bubble wall separates space–time into two parts, described by different metrics and containing different kinds of matter. The bubble wall is a timelike, spherically symmetric hypersurface Σ, the interior of the bubble is described by a de Sitter space–time and the exterior by the static universe discussed in Sect. 8.2. The Israel junction conditions [39] are implement in order to joint these two manifolds along there common boundary Σ. The evolution of the bubble wall is determined by implement these conditions. Unit as such that $8\pi G = 1$. The exterior of the bubble is described by the metric Eq. (8.1) and the Eqs. (8.2–8.5), previously discussed in Sect. 8.2. At the end, the static solution for these equations will be assumed. The interior of the bubble will be described by the metric of the de Sitter space–time in its open foliation, see [34]

$$ds^2 = dT^2 - b^2(T)\left(\frac{dz^2}{1+z^2} + z^2\, d\Omega_2\right), \tag{8.21}$$

where the scale factor satisfies

$$\left(\frac{db}{dT}\right)^2 = \left(\frac{V_T}{3}\right) b^2(T) + 1. \tag{8.22}$$

These two regions are separated by the bubble wall Σ, which will be assumed to be a thin-shell and spherically symmetric. Then, the intrinsic metric on the shell is [40]

$$ds^2|_{\Sigma} = d\tau^2 - B^2(\tau)\, d\Omega_2, \tag{8.23}$$

where τ is the shell proper time.

Now we proceed to impose the Israel matching conditions [39] in order to joint the manifolds along there common boundary Σ. The first of Israel's conditions impose that the metric induced on the shell from the bulk 4-metrics on either side should match, and be equal to the 3-metric on the shell. Then by looking from the outside to the bubble-shell we can parameterize the coordinates $r = x(\tau)$ and $t = t(\tau)$, obtaining the following match conditions, see [41]

$$a(t)x = B(\tau), \qquad \left(\frac{dt}{d\tau}\right)^2 = 1 + \frac{a(t)^2}{1-\left(\frac{x}{R}\right)^2}\left(\frac{dx}{d\tau}\right)^2, \tag{8.24}$$

where all the variables in these equations are thought as functions of τ. On the other hand, the angular coordinates of metrics (8.1) and (8.23) can be just identified in virtue of the spherical symmetry.

The second junction condition could be written as follow

$$[K_{ab}] - h_{ab}[K] = S_{ab}, \tag{8.25}$$

where K_{ab} is the extrinsic curvature of the surface Σ and square brackets stand for discontinuities across the shell. Following [41], we assume that the surface energy-momentum tensor S_{ab} has a perfect fluid form given by $S_\tau{}^\tau \equiv \sigma$ and $S_\theta{}^\theta = S_\phi{}^\phi \equiv -\bar{P}$, where $\bar{P} = (\bar{\gamma} - 1)\sigma$.

In the outside coordinates we parameterize $x(t)$ as the curve for the bubble evolution (the bubble radius in these coordinates). Since x and t are dependent variables on the shell, this is legitimate.

Then, from the Israel conditions we can obtain the following equation for the evolution of $x(t)$ and $\sigma(t)$, see [1]

$$\frac{dx}{dt} = \pm\sqrt{\frac{\left(R^2 - x^2\right)\left(a_0^2\, C^2\, x^2 - 1\right)}{x^2\, a_0^2\left(a_0^2\, C^2\, R^2 - 1\right)}}, \tag{8.26}$$

$$\frac{d\sigma}{dt} = -2\left(\frac{\bar{\gamma}\sigma}{x}\right)\frac{dx}{dt} + \frac{a_0\gamma\rho_0}{\sqrt{-\left(\frac{dx}{dt}\right)^2 a_0^2 + 1 - \frac{x^2}{R^2}}}\frac{dx}{dt}. \tag{8.27}$$

Where

$$C^2 = \frac{V_T}{3} + \left(\frac{\sigma}{4} + \frac{1}{\sigma}\left[\frac{V_F - V_T}{3} + \frac{A}{3a^{3\gamma}}\right]\right)^2. \tag{8.28}$$

The positive energy condition $\sigma > 0$ together with Israel conditions impose the following restriction to σ

$$0 < \sigma \leq 2\sqrt{\frac{V_F - V_T}{3} + \frac{\rho_0}{3}}. \tag{8.29}$$

Also, from the definition of x and Eq. (8.26) we obtain the following restriction for x

$$\frac{1}{a_0 C} \leq x \leq R. \tag{8.30}$$

We solved the Eqs. (8.26, 8.27) numerically by consider different kind and combinations of the matter content of the background and the bubble wall. From these solutions we found that once the bubble has materialized in the background of an ES universe, it grows filling completely the background space.

In order to find the numerical solutions we chose the following values for the free parameters of the model, in units where $8\pi G = 1$:

$$a_0 = 1, \tag{8.31}$$

$$V_T = 0.1 V_F, \tag{8.32}$$

$$\sigma_{init} = 10^{-6}. \tag{8.33}$$

The other parameters are fixed by the conditions discussed in Sect. 8.2.

Some of the numerical solutions are shown in (Fig. 8.4) where the evolution of the bubble, as seen by the outside observer, is illustrated. In these numerical solutions

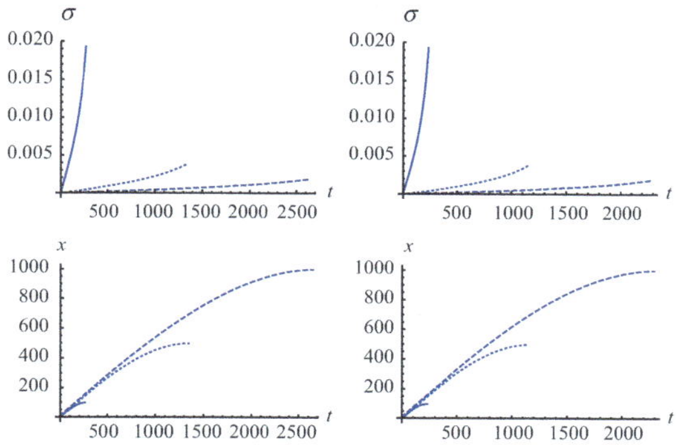

Fig. 8.4 Time evolution of the bubble in the outside coordinates $x(t)$, and time evolution of the surface energy density $\sigma(t)$. The *left panel* is for a static universe dominated by dust and the bubble wall containing dust. The *right panel* is the same situation but with radiations instead of dust. In all these graphics we have considered dashed line for $R = 1000$, *dotted line* for $R = 500$ and continuous line for $R = 100$

we have considered three different curvature radius ($R = 1000$, $R = 500$, $R = 100$) and various matter contents combinations for the background and the bubble wall.

From these examples we can note that the bubble of the new face grows to fill the background space, where the shell coordinate asymptotically tends to the curvature radius R.

8.5 Conclusions

In this work we explore an alternative scheme for an Emergent Universe scenario developed in [1], where the universe is initially in a truly static state. This state is supported by a scalar field which is located in a false vacuum. The universe begins to evolve when, by quantum tunneling, the scalar field decays into a state of true vacuum.

In particular, we study the process of tunneling of a scalar field from the false vacuum to the true vacuum and the consequent creation and evolution of a bubble of true vacuum in the background of Einstein static universe. The motivation in doing this is because we are interested in the study of new ways of leaving the static period and begin the inflationary regime in the context of Emergent Universe models.

In the first part, we study a Einstein static universe dominated by two fluids, one is a standard perfect fluid and the other is a scalar field located in a false vacuum. The requisites for obtain a static universe under these conditions are discussed. In the second part, we study the tunneling process of the scalar field from the false vacuum to the true vacuum and the consequent creation of a bubble of true vacuum in the background of Einstein static universe. Following the formalism presented in [35] we found the semiclassical tunneling rate for the nucleation of the bubble in this curved space. We conclude that the probability for the bubble nucleation is enhanced by the effect of the curvature of the closed static universe background. In the third part of the paper, we study the evolution of the bubble after its materialization. By following the formalism developed by Israel [39] we found that once the bubble has materialized in the background of an ES universe, it grows filling completely the background space. In particular, we use the approach of [41] to find the equations which govern the evolution of the bubble in the background of the ES universe. These equations are solved numerically, some of these solutions, concerning several type of matter combinations for the background and the bubble wall, are shown in (Fig. 8.4).

In resume we have found that this new mechanism for an Emergent Universe is plausible and could be an interesting alternative to the realization of the Emergent Universe scenario.

We have postpone for future work the study of this mechanism applied to Emergent Universe based on alternative theories to General Relativity, like Jordan–Brans–Dicke [42], which present stable past eternal static regime [16, 17]. It is interesting explore this possibility because Emergent Universe models based on GR suffer from instabilities, associated with the instability of the Einstein static universe [15]. This instability is possible to cure by going away from GR, for example, by consider a Jordan Brans Dicke theory, see [16, 17]. Another possibility is considering non-perturbative quantum corrections of the Einstein field equations, either coming from a semiclassical state in the framework of loop quantum gravity [9, 13] or braneworld cosmology with a timelike extra dimension [12, 14]. In addition to this, consideration of the Starobinsky model, exotic matter [10, 11] or the so-called two measures field theories [18–21] also can provide a stable initial state for the emergent universe scenario.

In the context of GR the instability of the ES could be overcome by consider a static universe filled with a non-interacting mixture of isotropic radiation and a ghost scalar field [43] or by consider a negative cosmological constant with a universe dominated by a exotic fluid satisfies $P = (\gamma - 1)\rho$ with $0 < \gamma < 2/3$, see [44].

Acknowledgements This work has been partially supported by FONDECYT grant N^0 11090410, Mecesup UBB0704 and Universidad del Bío-Bío through grant DIUBB 121407 GI/VC.

References

1. Labrana, P. "Emergent Universe by Tunneling," Phys. Rev. D **86**, 083524 (2012) [arXiv:1111.5360 [gr-qc]]
2. Borde A. and Vilenkin A., Eternal inflation and the initial singularity, 1994 Phys. Rev. Lett. **72** 3305
3. Borde A. and Vilenkin A., Violation of the weak energy condition in inflating spacetimes, 1997 Phys. Rev. D **56** 717
4. Guth, A.H., Eternal inflation, arXiv:astro-ph/0101507
5. Borde A., Guth A. H. and Vilenkin A., Inflationary space-times are incompletein past directions, 2003 Phys. Rev. Lett. **90** 151301
6. Vilenkin A., Quantum cosmology and eternal inflation, arXiv:gr-qc/0204061

7. Ellis G. F. R. and Maartens R., The emergent universe: Inflationary cosmology with no singularity, 2004 Class. Quant. Grav. **21** 223
8. Ellis, G.F.R., Murugan J. and Tsagas C. G., The emergent universe: An explicit construction, 2004 Class. Quant. Grav. **21** 233
9. Mulryne D. J., Tavakol R., Lidsey J. E. and Ellis G. F. R., An emergent universe from a loop, 2005 Phys. Rev. D **71** 123512
10. Mukherjee S., Paul B. C., Maharaj S. D. and Beesham A., Emergent universe in Starobinsky model, arXiv:gr-qc/0505103
11. Mukherjee S., Paul B. C., Dadhich N. K., Maharaj S. D. and Beesham A., Emergent universe with exotic matter, 2006 Class. Quant. Grav. **23** 6927
12. Banerjee A., Bandyopadhyay T. and Chakraborty S., Emergent universe in brane world scenario, arXiv: 0705.3933 [gr-qc]
13. Nunes, N.J., Inflation: A graceful entrance from loop quantum cosmology, 2005 Phys. Rev. D **72** 103510
14. Lidsey J. E. and Mulryne D. J., A graceful entrance to braneworld inflation, 2006 Phys. Rev. D **73** 083508
15. A. S. Eddington, Mon. Not. Roy. Astron. Soc. **90**, 668 (1930)
16. S. del Campo, R. Herrera and P. Labrana, JCAP **0711** 030 (2007)
17. del Campo, S., R. Herrera, P. Labrana, JCAP **0907**, 006 (2009). [arXiv:0905.0614 [gr-qc]]
18. del Campo, S., Guendelman, E., R. Herrera, P. Labrana, JCAP **1006** (2010) 026. [arXiv:1006.5734 [astro-ph.CO]]
19. del Campo, S., Guendelman, E.I., Kaganovich, A.B., R. Herrera, P. Labrana, Phys. Lett. **B699** (2011) 211-216. [arXiv:1105.0651 [astro-ph.CO]]
20. E.I. Guendelman, [arXiv:1103.1427 [gr-qc]]
21. E.I. Guendelman, [arXiv:1105.3312 [gr-qc]]
22. Banerjee, A., T. Bandyopadhyay, S. Chakraborty, Gen. Rel. Grav. **40**, 1603-1607 (2008). [arXiv:0711.4188 [gr-qc]]
23. U. Debnath, Class. Quant. Grav. **25**, 205019 (2008). [arXiv:0808.2379 [gr-qc]]
24. B. C. Paul, S. Ghose, Gen. Rel. Grav. **42**, 795-812 (2010). [arXiv:0809.4131 [hep-th]]
25. Beesham, A., Chervon, S.V., S.D. Maharaj, Class. Quant. Grav. **26**, 075017 (2009). [arXiv:0904.0773 [gr-qc]]
26. U. Debnath, S. Chakraborty, Int. J. Theor. Phys. **50**, 2892-2898 (2011). [arXiv:1104.1673 [gr-qc]]
27. Mukerji, S., Mazumder, N., R. Biswas, S. Chakraborty, Int. J. Theor. Phys. **50**, 2708-2719 (2011). [arXiv:1106.1743 [gr-qc]]
28. A. Linde, Phys. Rev. D **59**, 023503 (1998)
29. A. Linde, M. Sasaki and T., Phys. Rev. D **59**, 123522 (1999)
30. S. del Campo and R., Phys. Rev. D **67**, 063507 (2003)
31. del Campo, S., Herrera, R., Saavedra, J., Phys. Rev. D **70**, 023507 (2004)
32. Balart, L., del Campo, S., Herrera, R., Labrana, P., Saavedra, J., Phys. Lett. **B647**, 313-319 (2007)
33. S. R. Coleman, Phys. Rev. D **15**, 2929 (1977) [Erratum-ibid. D **16**, 1248 (1977)]
34. S. R. Coleman and F. De Luccia, Phys. Rev. D **21**, 3305 (1980)
35. E. Keski-Vakkuri and P. Kraus, Phys. Rev. D **54**, 7407 (1996) [arXiv:hep-th/9604151]
36. Basu, R., A. H. Guth, A. Vilenkin, Phys. Rev. **D44**, 340-351 (1991)
37. D. Simon, J. Adamek, A. Rakic and J. C. Niemeyer, JCAP **0911**, 008 (2009) [arXiv:0908.2757 [gr-qc]]
38. Abbott, L.F., Harari, D., Q.H. Park, Class. Quant. Grav. **4**, L201 (1987)
39. W. Israel, Nuovo Cim. B **44S10**, 1 (1966) [Erratum-ibid. B **48**, 463 (1967)] [Nuovo Cim. B **44**, 1 (1966)]
40. Berezin, V.A., Kuzmin, V.A., I.I. Tkachev, Phys. Rev. D **36**, 2919 (1987)
41. Fischler, W., Paban, S., M. Zanic, C. Krishnan, JHEP **0805**, 041 (2008). [arXiv:0711.3417 [hep-th]]

42. Jordan P., The present state of Dirac's cosmological hypothesis, 1959 Z.Phys. **157** 112; Brans C. and Dicke R. H., Mach's principle and a relativistic theory of gravitation, 1961 Phys. Rev. **124** 925
43. Barrow, J.D., C.G. Tsagas, Class. Quant. Grav. **26**, 195003 (2009)
44. P. W. Graham, B. Horn, S. Kachru, S. Rajendran and G. Torroba, arXiv:1109.0282 [hep-th]

Chapter 9
A Review on the Scalar Field/Bose-Einstein Condensate Dark Matter Model

Abril Suárez, Victor H. Robles and Tonatiuh Matos

Abstract We review the work done so far aimed at modeling in an alternative way the dark matter in the Universe: the scalar field Bose-Einstein condensate dark matter (SFDM/BEC) model. We discuss a number of important achievements and characteristics of the model. We also describe some of our most recent results and predictions of the model compared to those of the standard model of ΛCDM.

9.1 Introduction

It is a pleasure for us to review the different theoretical basis of the SFDM/BEC model as a dark matter (DM) candidate. We think this review is important for mainly two reasons: the considerable progress in the model since it was proposed as a serious candidate to the dark matter paradigm, improved theoretical understanding of the nature of the DM and the significant advances in the cosmological and astronomical observations are leading us to put more constraints and will allow us to test the model and decide if it can still stand as a viable DM paradigm or if it should be discarded.

Recent observations of the Universe have found that only 4 % of the total content of the Universe is baryonic matter, being 22 % of the rest remaining non-baryonic dark matter (DM) and the rest in some form of cosmological constant.

The incorporation of a new kind of DM different from that proposed by the standard model, also known by Λ-Cold Dark Matter model (ΛCDM), into the Big Bang Theory holds out the possibility of giving alternative answers to some of the unsolved issues of the standard cosmological model. Several authors have proposed interesting alternatives in where they try to solve the difficulties that the ΛCDM scenario seems not to solve. In fact, the alternative Scalar Field Dark Matter (SFDM/BEC)

A. Suárez (✉) · V. H. Robles · T. Matos
Departamento de Física, Centro de Investigación y de Estudios Avanzados del I. P. N., Apdo. 14–740, D.F., México,
e-mail: asuarez@fis.cinvestav.mx

V. H. Robles
e-mail: vrobles@fis.cinvestav.mx

T. Matos
e-mail: tmatos@fis.cinvestav.mx

C. Moreno González et al. (eds.), *Accelerated Cosmic Expansion*, Astrophysics and Space Science Proceedings 38,
DOI 10.1007/978-3-319-02063-1_9,

scenario has received much attention in the last few years. The main idea is simple, in this models the nature of the DM is completely determined by a fundamental scalar field Φ [45].

The idea was first considered in [59] and independently introduced in [45, 46, 65] suggesting bosonic dark matter as a model for galactic halos, see also [33]. In the SFDM/BEC model, DM halos can be described, in the non-relativistic regime, as Newtonian gravitational condensates made up of ultra-light bosons that condense into a single macroscopic wave function.

Several authors have introduced a dynamical scalar field with a certain potential $V(\Phi)$ as a candidate to be the dark matter, although there is not yet an agreement for the correct form of the potential field. Other interesting works consider a single scalar field to unify the description of dark matter, dark energy and inflation [55, 72, 73].

Different issues of the cosmological behavior of the SFDM/BEC model have been studied in a wide variety of approaches, see for example [3, 6, 26, 51, 53, 56, 61, 77, 79, 82–84, 110, 111, 119, 123, 125]. For example, [56] proposed fuzzy dark matter composed of ultra-light scalar particles initially in the form of a BEC. Recently [51, 53] developed a further analysis of the cosmological dynamics of SFDM/BEC as well as the evolution of their fluctuations (see also [26]). In the same direction, [78, 119] studied the growth of scalar fluctuations and the formations of large-scale structure within a fluid and a field approach for the SFDM/BEC model.

In addition, many numerical simulations have been performed to study the gravitational collapse of the SFDM/BEC model [13–15, 28, 38, 47–49, 90, 92]. [24, 25] found an approximate analytical expression and numerical solutions of the mass-radius relation of SFDM/BEC halos. Recently, [106] gave constraints on the boson mass to form and maintain more than one vortex in SFDM/BEC halos. These constraints are in agreement with the ultra-light mass found in previous works (see for example, [60]). Lately, [75] performed N-body simulations to study the dynamics observed in the Ursa Minor dwarf galaxy. They modeled the dark matter halo of Ursa Minor as a SFDM/BEC halo to establish constraints for the bosons mass. Moreover, they introduced a dynamical friction analysis within the SFDM/BEC model to study the wide distribution of globular clusters in Fornax. An overall good agreement is found for the ultra-light mass of bosonic dark matter.

In this paper we review a number or results and important characteristics of the SFDM/BEC model, its dynamical mechanism and some of its predictions. We also discuss some of the trending topics nowadays in the subject which attempt to predict and ask how well the model is achieving its goals. As we will see, a number of them studies results which are in reasonable agreement with the general features required by the theory and the data.

The outline of the paper is as follows. In Sect. 1 we have given a brief synopsis on the current state-of-the-art of the model. In the Sect. 2 we describe the dark matter paradigm and briefly resume the standard model of cosmology. In Sect. 3 and 4 we describe in some detail why Scalar Field/Bose–Einstein condensate Dark Matter (SFDM/BEC) can be a good alternative candidate to be the dark matter in our Universe, we summarize some representative papers for this sections, and in Sect. 5 we include topics for future works and our conclusions.

9.2 The Dark Matter Paradigm and the Standard Model of Cosmology

In this millennium, new technologies are opening wider windows to explore our Universe. For some time we could only relay on inaccurate evidence found in the local neighborhood of our galaxy to infer the history of our Universe, now it turns it is possible for us to see the evolution of the Universe as far as 100,000 years after the Big Bang and in more detail.

With these advances, nowadays some inquires of our cosmic evolution can be determined by giving an answer to question like: How much matter is in the Universe?

Since the discovery of the expansion of the Universe done by Hubble and Slipher [57] in the 1920's, the common believe had been that all energy in the Universe was in the form of radiation and ordinary matter (electrons, protons, neutrons, etc.). Over the past few decades, theories concerning the stability of galaxies ([9, 112]) indicated that most of the mass in our Universe is dark (i.e., it does not emits or absorbs light [93, 121]), therefore resulting unobservable by telescopes. The suggestion that "dark matter" may form a large fraction of the density in the Universe was raised by Zwicky in 1937. Back then he used the virial theorem to obtain the average mass of galaxies within the Coma cluster and obtained a value much larger than the mass of the luminous material, he then realized that some mass was "missing" in order to account for observations. This missing mass problem was confirmed many years later by more accurate measurements of rotation curves of disc galaxies, [17, 19, 23, 99, 112]. The rotation curves of neutral hydrogen clouds in spiral galaxies measured by the Doppler effect are now found to be roughly flat with a typical rotation velocity equal to $v_\infty \sim 200$ km/s up to the maximum observed radius of about 50 kpc. With these observations the mass profile results much more extended than the common distributions which typically converge within ~ 10 kpc. This would imply that galaxies might be surrounded by an extended halo of dark matter whose mass $M(r) \sim rv_\infty^2/G$ increases linearly with radius (here r is the radius and G Newton's gravitational constant).

In the 1980's, the proposal of dark matter found its basis in the so called "inflationary scenario" [1, 42, 74], a theory of the first 10^{-30} s developed to give answer to several questions left unanswered by the Big Bang model, like for example: Why is the Universe so homogeneous and isotropic? and; Where did the initial homogeneities that gave rise to the structures we see today came from? [12, 43, 44, 54, 118]. The inflationary theory predicts that the Universe is spatially flat; which according to Einstein's theory of general relativity, this fixes the total energy density of the Universe making it equal to the critical value, $\rho_c \equiv 3H_0^2/8\pi G \sim 1.7 \times 10^{-29}$g cm^3, where H_0 is the current value of the Hubble parameter.

Several astrophysical observations of distant type Ia supernovae have also shown that the content of the Universe is made of about 74 % of dark energy, 22 % of dark matter and 4 % of baryonic (visible) matter, [22, 31, 34, 35, 50, 100, 103, 115, 126]. Inflation thus seemed to call for dark matter.

It then results that most matter in the Universe is non-luminous. The observed flatness of the galactic rotation curved indicating the presence of dark matter halos around galaxies [112]. Summed to these evidences are the observations of the cosmic microwave background (CMB) anisotropies [58, 64, 117] combined with large-scale structure and type Ia supernova luminosity data [100, 103–105] which all together constrain the cosmological parameters also finding once more that visible matter contributes only about 4 % of the energy density of the Universe, gravitational lensing [102] and X-ray spectra [36, 37, 71, 89, 121] in elliptical galaxies, and the high velocity dispersion and gas temperature in clusters of galaxies [76, 121], all of them leading to a picture in which galaxies are composed of a luminous galactic disk surrounded by a galactic halo of dark matter. Also the relative contribution of the dark matter component is usually specified in terms of the mass-to-light ratio, M/L; which reflects the total amount of mass relative to the total light within a given scale. The increase on this ratio suggests that there is relatively more dark than luminous matter with increasing scale [112]. This has led to the general belief that clusters have more dark matter per unit luminosity than individual galaxies and that superclusters may have even more. This widely accepted monotonic increase of M/L with scale determines to a large extent the prevalent views about the location of the dark matter and the total mass density of the Universe. Recent studies of the dependence of the mass-to-light ratio on scale indicate that M/L is nearly constant on large scales ranging up to supercluster size (10 Mpc), suggesting no additional dark matter is tucked away on large scales [10]. More recently, a clear separation between the center of baryonic matter and the total center of mass was observed in the Bullet cluster [27] and later in other galaxy cluster collisions [20].

The observational evidence for dark matter continues to grow, and particle physicists have proposed various particles, motivated by supersymmetry and unified theories, that could reasonably explain it. These observations reinforce the claim that dark matter is indeed composed of weakly interacting particles and is not a modification of gravity. However, even taking into account all these results the properties of dark matter are still mysterious.

It then also results that an important question in cosmology has to do with knowing the nature of the so far undefined one quarter part of the content of the Universe, the dark matter. As mentioned before, the cosmological observations seem to support the idea that dark matter can be made of some kind of non-baryonic, non-relativistic and weakly interactive massive particle. Many efforts trying to give an answer to this question have been done in the past few decades, mainly motivated by the idea that the answer will probably change our understanding of the Universe and its dynamics. One of the explanations for DM is the SFDM model.

In the Standard Model of cosmology, the total energy density of the Universe is dominated today by the densities of two components: the "dark matter" which has an attractive gravitational effect like usual matter and the "dark energy" which can be considered as a kind of vacuum energy with a negative pressure, which seems constant today (i.e., a cosmological constant, Λ). Although the real nature of these two components remains unknown, in the standard model dark matter is generally modeled as a system of collisionless particles. This is known as the "Λ Cold Dark Matter" model, which predicts that the Universe contains primarily cold neutral weakly interactive massive particles (WIMPs) which are non-baryonic [95, 128], pressureless, behave like a cold gas, one beyond those existing in the Standard Model of particle physics and have not yet been detected in accelerators or specialized indirect searches, in particular, the lightest supersymmetric particles, the most popular of which is the neutralino, with a particle mass of the order of 100 GeV. Efforts are underway to measure the presence of these particles, but no direct detection has yet been reported.

In order to explain observational data, the ΛCDM model was developed, [94, 96] and it is the most simple possibility. In the Standard Model of cosmology the matter component $\Omega_M \sim 26\,\%$ of the Universe decomposes itself into baryons, neutrinos, etc., and cold dark matter which would be responsible for the formation of structure in the Universe. Observations indicate that stars and dust (baryons) represent something like 0.4 % of the total content of matter in the Universe. The measurements of neutrino masses indicate that these contribute nearly with the same amount as matter. In other words, $\Omega_M = \Omega_m + \Omega_{DM} = \Omega_b + \Omega_\nu + \cdots + \Omega_{DM} \sim \Omega_m + \Omega_{CDM}$, where Ω_{CDM} represents the cold dark matter part of the matter contributions, and has a value of $\Omega_{CDM} \sim 0.22$. The value of the amount of baryonic matter is in accordance with the limits imposed by nucleosynthesis, [114]. This model then considers a flat Universe ($\Omega_\Lambda + \Omega_M \equiv 1$) with 96 % of unknown matter but which is of great importance in the cosmological context. It also supposes a homogeneous and isotropic Universe which evolution can be best described today by Friedmann's equations coming from general relativity and whose main ingredients can be described by fluids with characteristics very similar to those we see in our Universe. We now know that the Universe is not exactly homogeneous and isotropic, but the standard model does give a framework within which the evolution of structures such as galaxies or clusters of galaxies can be studied with their origins coming from small fluctuations in the density of the early Universe. The model assumes a "scale-invariant" spectrum of initial density fluctuations, a spectrum in which the magnitude of the inhomogeneity is the same on all length scales, again as predicted by standard inflationary cosmology [12, 43, 118, 126]. Moreover, ΛCDM seems to be until today the most successful model fitting current cosmological observations [11].

The ΛCDM model successfully describes the accelerated expansion of the Universe, it explains the Cosmic Microwave Background radiation in great detail and provides a framework within which one can understand the large-scale isotropy of the Universe, it also describes the important characteristics of the origin, nature and evolution of the density fluctuations which are believed to give rise to galaxies and other cosmic structures, the Lyman-α forest, the large scale matter distribution, and the main aspects of the formation an the evolution of virialized cosmological objects. So far the ΛCDM model is consistent with the observed cluster abundance at $z \sim 0$, it then predicts a relatively little change in the number density of rich clusters as a function of redshift because, due to the low matter density, hardly any structure growth has occurred since $z \sim 1$. The ΛCDM model can then be "forced" to agree approximately with both the cluster abundance on small scales and the CMB fluctuations on large scales by tilting the power spectrum (by about 30 %) from its standard shape. This tilted variant of the ΛCDM model then results nearly consistent with observations. The power spectra of ΛCDM can then be normalized so that it agrees with both the CMB and cluster observations. But as the estimates of the cold dark matter density become more precise, it becomes even more imperative for its composition to be identified.

There remain, however, certain conflicts at galactic scales, like the cusp profile of central densities in galactic halos, the overpopulation of substructures predicted by N-body numerical simulations which are an order of magnitude larger than what has been observed, among others, see for example [27, 63, 88, 98]. And until today the nature of the dark matter that binds galaxies remains an open question.

9.3 Why Scalar Field Dark Matter?

In the big bang model, gravity plays an essential role: it collects the dark matter in concentrated regions called 'dark matter halos'. Within these large dark matter halos, the baryons are believed to be so dense that they radiate enough energy to collapse into galaxies and stars. The most massive halos, hosts for the brightest galaxies, are formed in regions with the highest local mass density. Less massive halos, hosts for the less bright galaxies, appear in regions with low local densities [97]. These situations appear to be the same as in our extragalactic neighborhood, but there are still problems. Despite all its successful achievements the ΛCDM model requires further considerations.

The ΛCDM paradigm faces several challenges to explain observations at galactic scales, such as the central densities of dark halos, dwarfs and Low Surface Brightness (LSB) galaxies, the excess of satellite galaxies predicted by N-body simulations, the formation of bars in disc galaxies, etc. [27, 88, 98]. In other words, there is not a match between ΛCDM predictions at galactic scales and what is being observed. Problems with an otherwise successful model are often the key to a new and deeper understanding.

Observations point out to a better understanding of the theory beginning with the Local Void, which contains just a few galaxies that are larger than expected. This problem would be solved if structure grew faster than in the standard theory, therefore filling the local void and giving rise to more matter in the surroundings [97].

Another problem arises for the so called pure disk galaxies, which do not appear in numerical simulations of structure formation in the Standard Model. These problems would be solved again if the structure grew faster than it does in the standard paradigm [97].

On the other hand, [68] also found that the collision velocity of 3000 km/s at R_{200} for the Bullet Cluster is very unlikely within the ΛCDM paradigm, which moves it to a challenge for the Standard Model of cosmology.

A final example of inconsistencies can be seen in a paper by [120], who found anomalies in the mass power spectrum obtained by the SDSS and the one obtained with the ΛCDM model, i.e., anomalies in the predicted large-scale structure of the

Universe. With these and other results it seems necessary to change the ΛCDM paradigm to try and explain the formation of structure in the Universe.

Given these discrepancies, it seems necessary to explore alternatives to the paradigm of structure formation. These are some of the reasons why we need to look for alternative candidates that can explain the structure formation at cosmological level, the observed amount of dwarf galaxies, and the dark matter density profiles in the core of galaxies. Recently, several alternative models have been proposed.

One of them invokes a scalar field as dark matter in the Universe [83, 84]. This model supposes that dark matter is a real or complex scalar field Φ minimally coupled to gravity, endowed with a scalar potential $V(\Phi)$ and that at some temperature it only interacts gravitationally with the rest of the matter. This scalar field can be added to the particles standard model lagrangian or to the general relativity one, supposing that the coupling constant with the rest of the matter is very small. It has been also suggested that this scalar field can be derived from higher dimensional theories. It has also been proposed that this dark matter scalar field, i.e., this spin-0 fundamental interaction, could lead to the formation of Bose-Einstein condensate in the way of cosmic structure [33, 51, 53, 56, 82] with an ultra-light mass of order $m \sim 10^{-22}$ eV. From this mass it follows that the critical temperature of condensation $T_c \sim 1/m^{5/3\sim}$ TeV is very high, therefore, they may form Bose-Einstein condensate drops very early in the Universe [83] that behave as cold DM. Lee and Koh [65], and independently Matos and Guzmán [45], suggested bosonic dark matter as a model for galactic halos. In addition, the Compton length $\lambda_c = 2\pi\hbar/m$ associated to this boson results of about ~kpc, and corresponds to the dark halo size of typical galaxies in the Universe. Thus, it has been supported that these drops are the halos of galaxies (see [82]), i.e., that halos are huge drops of SF. In a recent paper, Ureña [123] studied the conditions for the formation of a SFDM/BEC in the Universe, also concluding that SFDM/BEC particles must be ultra-light bosons.

In the SFDM model the initial halos of galaxies do not form hierarchically, they are formed at the same time and in the same way when the Universe reaches the critical temperature of condensation of the SF. From this it follows that galaxies can share some properties because they formed in the same manner and at the same moment [69]. Therefore, from this paradigm we have to expect that there exists well formed galaxy halos at higher redshifts than in the ΛCDM model. Recently Suárez and Matos [119] developed a hydrodynamical approach for the structure formation in the Universe with the scalar potential $V(\Phi) = m^2\Phi^2/2 + \lambda\Phi^4/4$. They found that when $\lambda = 0$ the evolution of perturbations of the SFDM model compared to those of ΛCDM are identical. They also showed that this potential can lead to the early formation of gravitational structures in the Universe depending on the sign of the self-interaction parameter λ.

The most simple model having both an exponential behavior and a minimum is a *cosh*-like potential. Another interesting work was done by [80] an independently by [113] who used a potential of the form $V(\Phi) = V_0[cosh(\eta\Phi - 1)]$ where V_0 and η are constants to explain the core density problem for disc galaxy halos in the ΛCDM model (see also [29, 101]) and to perform the fist cosmological analysis in the context of SFDM. They showed that the evolution of the Universe, its expansion rate and the growth of linear perturbations in this model are identical as those derived in the standard model. In [45] they model the dark matter in spiral galaxies, assuming dark matter as an arbitrary scalar field endowed with a scalar potential.

Another scalar potential widely used to describe dark matter is $V(\Phi) = m^2\Phi^2/2$ [83, 122]. This potential is very interesting because it can mimic the cosmological evolution of the Universe predicted by the ΛCDM model. Also, it is known that an exponential-like scalar field potential fits very well the cosmological constraints due to the form of its solutions (see for example [29, 101, 127]). If the self-interaction of the SF is considered, we need to add a quartic term to the SF potential [3, 6–8, 21], in this case the equation of state of the SF results to be that of a polytrope of index $n = 1$ (see [4, 51, 53, 70, 119]).

Another interesting result is that the predicted density of neutrinos at the recombination epoch is in agreement with the observations of the Wilkinson Microwave Anisotropy Probe (WMAP). In the same direction, Rodríguez–Montoya et al. [110] studied ultra-light bosons as dark matter in the Universe with the framework of kinetic theory, through the Boltzmann–Einstein equations, and they found that this kind of ultra-light particles is consistent with the acoustic peaks of the cosmic microwave background radiation if the boson mass is around $m \sim 10^{-22}$ eV.

[66] pointed out that SFDM/BEC can explain the spatial separation of the dark matter from visible matter, as derived from X-ray maps and weak gravitational lensing, in the Bullet Cluster, see also [27].

Other works have used the bosonic dark matter model to explain the structure formation via high-resolution simulations. [123, 124] reviewed the key properties that may arise from the bosonic nature of SFDM models. On the other hand, several authors have numerically studied the formation, collapse and viralization of SFDM/BEC halos as well as the dynamics of the SFDM around black holes [13, 25, 30, 39, 47–49, 129]. In [2], Alcubierre et al. found that the critical mass for collapse of the SF is of the order of a Milky Way-sized halo mass. This suggests that SFDM/BEC can be plausible candidate to dark matter in galactic halos. In addition, Lora et al. [75], studied, through N-body simulations, the dynamics of Ursa Minor dwarf galaxy and its stellar clump assuming a SFDM/BEC halo to establish constraints for the boson mass. Moreover, they introduced a dynamical friction analysis with the SFDM/BEC model to study the distribution of globular clusters in Fornax. An overall good agreement is found for the ultra-light mass $\sim 10^{-22}$ eV of bosonic dark matter.

In this model the scalar particles with ultra-light mass are such that their wave properties avoid the cusp problem and reduce the high number of small satellites by the quantum uncertainty principle [52, 56, 63, 77, 107]. Robles and Matos [107], (see also [16, 18, 52]) showed that BEC dark matter halos fit very well high-resolution rotation curves of LSB galaxies, and that the constant density core in dark halos can be reproduced. Also, [69] showed how the SFDM/BEC paradigm is a good alternative to explain the common mass of the dark halos of dwarf spheroidal galaxies. Recently, Rindler–Daller and Shapiro, [106], investigated the formation of vortex in SFDM/BEC halos. They found constraints on the boson mass in agreement with the ultra-light mass found in previous works (see also [60, 87, 130]), some of these issues will be discussed in more detail in Sect. 4.

Summarizing, it is remarkable that with only one free parameter, the ultra-light scalar field mass ($m \sim 10^{-22}$ eV), the SFDM model fits:

i) The evolution of the cosmological densities [83].
ii) The central density profile of the dark matter is flat [16].
iii) The acoustic peaks of the cosmic microwave background [110].
iv) The scalar field has a natural cut off, thus the substructures in cluster galaxies are suppressed naturally. With a scalar field mass of $m_{\Phi} \sim 10^{-22}$ eV the amount of substructures is compatible with the ones observed [82, 56, 119].
v) We expect that SFDM forms galaxies earlier than the ΛCDM model, because they form BEC's at a critical temperature $T_c \gg M$ eV. So if the SFDM model is right, we have to see big galaxies at high redshifts with similar features [83, 119].
vi) Adding self-interaction and Temperature correcctions, the rotation curves of big galaxies and LSB galaxies [5, 16, 53, 70, 108].
vii) With this mass, the critical mass of collapse for a real scalar field is just $10^{12} M_{\odot}$, i. e., the observed in galaxy halos [2].
vii) The observed properties of dwarf galaxies, i. e., the minimum length scale, the minimum mass scale, and their independence from the brightness [69].
ix) And recently it has been demonstrated that the SFDM halos would have cores large enough to explain the longevity of the cold clump in Ursa Minor and the wide distribution of globular clusters in Fornax [75].

Then, the SFDM/BEC model has provided to be a good candidate for dark matter halos of galaxies in the Universe because it can explain many aspects where the standard model of cosmology fails ([16, 18, 52, 66, 67, 69, 75, 107]). Therefore, not only the many successful predictions of the Standard Model of cosmology at large scales are well reproduced by SFDM, but also the ones at galactic scales. The scalar field models presents some advantages over the standard ΛCDM model like the ones mentioned above. Also, its self-interaction can, in principle, explain the smoothness of the energy density profile in the core of galaxies [82, 86]. Nevertheless, its important to remark that when a new dark matter candidate is proposed the study of the final object that will be formed as a result of a gravitational collapse is always an important but difficult task that requires continuos work since the baryonic physics is still poorly understood.

9.4 Current Status of the SFDM model

9.4.1 Self-Gravitating Bose-Einstein Condensate Dark Matter

Following [18], dark matter halos as a self-gravitating Bose-Einstein condensate with short-range interactions have been widely discussed [24, 25, 107, 108]. In these models, it is supposed that the cosmic BEC has a relatively low mean mass density so that the Newtonian approximation can be used. In the literature, when $T = 0$ all the bosons have condensed and the system can be described by one order parameter $\psi(\mathbf{r},t)$, called the condensate wave function. In the mean-field approximation, the ground state properties of the condensate are described by the Gross-Pitaevskii equation:

$$i\hbar\frac{\partial\psi}{\partial t}(\mathbf{r},t) = -\frac{\hbar^2}{2\,m}\Delta\psi(\mathbf{r},t) + m\Phi_{tot}(\mathbf{r},t)\psi(\mathbf{r},t).$$

where Φ_{tot} is the total potential exerted on the condensate. With this equation, [24] studied the structure and the stability of a self-gravitating BEC with short-range interactions. In this case, the results obtained in the absence of self-coupling and the results of [18] obtained for self-coupled BECs in the Thomas-Fermi approximation have been connected. The case of attractive short-range interactions were considered and the existence of a maximum mass above which no equilibrium state exists was found.

This study was motivated by the proposal that dark matter halos could be gigantic cosmic BEC's, [5, 18, 41, 56, 65]. In this case, gravitational collapse is prevented by the Heisenberg uncertainty principle or by the short-range interaction. This suggestion still remains highly speculative since the nature of dark matter remains unknown. On the other hand, whatever the nature of its constituents, if dark matter is viewed as a collisionless system described by the Vlasov equation, dark matter halos could result from processes of violent collisionless relaxation. In that case, gravitational collapse can be prevented by a Lynden-Bell type of exclusion principle, because this form of relaxation could be more rapid and efficient than a collisional relaxation. Furthermore, it generates a density profile with a flat core and a r^{-2} outer density profile for the halo, yielding flat rotation curves. These features are remarkably consistent with observations making this alternative scenario quite attractive.

In [25] the same author obtained the exact mass-radius relation of self-gravitating BECs with short-range interactions by numerically solving the equation of hydrostatic equilibrium taking into account quantum effects. He compared his results with the approximate analytical relation obtained in [24] from a Gaussian ansatz. He found that the Gaussian ansatz always provide a good qualitative agreement with the exact solution, and that the agreement is quantitatively very good.

In one of his most recent works Chavanis [26] assumed that the dark matter in the universe could be a self-gravitating BEC with short-range interactions, and he then theoretically explored the consequences of this hypothesis. He considered the possibilities of positive and negative scattering lengths.

At the level of dark matter halos a positive scattering length, equivalent to a repulsive self-interaction generating a positive pressure, is able to stabilize the halos with respect to gravitational collapse. This leads to dark matter halos without density cusps with an effective equation of state equal to that of a polytrope of index $n = 1$. Alternatively, if the scattering length is negative, equivalent to an attractive self-interaction generating a negative pressure, the dark matter is very unstable and collapses above a very small critical mass $M_{max} = 1.012\hbar/\sqrt{|a_s|Gm}$, where a_s is the scattering length. When these ideas were applied to an infinite homogeneous cosmic fluid, it was found that a negative scattering length can increase the maximum growth rate of the instability and accelerate the formation of structures. The virtues of these results could be combined by assuming that the scattering length changes sign in the course of the evolution. It could be initially negative to help with the formation of structures and become positive to prevent complete gravitational collapse. The mechanism behind this change of sign remains unknown. However, some terrestrial experiments have demonstrated that certain atoms can have negative scattering lengths and that it is possible in principle to manipulate the value and the sign of a_s.

It has also been found that a SFDM/BEC universe with positive scattering length, having a positive pressure, is not qualitatively very different from a classical Einstein-de Sitter universe. It also emerges at a primordial time $t = 0$ from a big-bang singularity where the density is infinite, and then undergoes a decelerating expansion. A difference, however, is that the initial scale factor $a(0)$ is finite. On the other hand, a SFDM/BEC universe with an always negative scattering length, having a negative pressure, markedly differs from previous models. It starts from $t \longrightarrow -\infty$ with a vanishing radius and a finite density, it has an initial accelerating expansion then decelerates and asymptotically behaves like the EdS universe. This model universe exists for any time in the past and there is no big-bang singularity. When the effect of radiation, baryonic matter and dark energy were added, the picture is quite different. In that case, a SFDM/BEC universe with attractive or repulsive self-interaction started from a singularity at $t = 0$ where the density was infinite. It first experiences a phase of decelerating expansion followed by a phase of accelerating expansion. For $k \longrightarrow 0$ ($k = 2\pi a_s \hbar^2/m^3c^2$ being a polytropic constant appearing the equation of state of the BEC and which depends on the scattering length a_s) the standard ΛCDM model was recovered but for $k \neq 0$, the evolution of the scale factor in a SFDM/BEC universe turned out to be substantially different. The model with $k > 0$ expands more rapidly than the standard model. The initial scale factor is finite and the radiation never dominates. The model with $k < 0$ expands less rapidly than the standard model. The initial scale factor vanishes and the radiation dominates leading to a decelerating expansion. In both models, the dark energy dominates at large times leading to an accelerating expansion. Finally, a dark fluid with generalized equation of state $p = (\alpha\rho + k\rho^2)c^2$ having a pressure component $p = k\rho^2c^2$ similar to a BEC dark matter and a component $p = \alpha\rho c^2$ mimicking the effect of the cosmological constant was considered. Optimal parameters (α, k) that gave a good agreement with the standard model were found. Also the growth of perturbations in these different models was studied and confirmed the previous observation of Harko, [53], that the density contrast increases more rapidly in a BEC universe than in the standard model.

In conclusion, it was pointed out that the idea that dark matter could be a BEC is fascinating and probably deserves further research.

9.4.2 BEC Dark Matter and Cosmological Perturbations

Through out this work we have shown how the SFDM/BEC model could be a serious alternative to the dark matter in the Universe. In [78] they studied in quite some detail the growth and virialization of Φ^2-dark matter perturbations in the linear and nonlinear regimes. Following the spherical collapse model, they also studied the nonlinear regime of the evolution of Φ^2-dark matter perturbations. They showed that the evolution of an overdense region of Φ^2-dark matter can collapse and virialize in a bound structure. However, they found that the scalar perturbations collapse at earlier times of the Universe than those in the CDM model. Thus, the standard and the SFDM/BEC model can be confronted in their predictions concerning the formation of the first galaxies. Massive galaxies at high redshifts is a prediction of the model and may be used to distinguish between SFDM/BEC paradigm and CDM.

As the study in [78] is detail in the analysis of perturbations we include a brief summary. In the study of the cosmological dynamics of the SFDM model it was considered the simplest case: a single scalar field $\tilde{\Phi}(x,t)$, with self-interacting double-well potential. They used the potential

$$V(\tilde{\Phi}) = \frac{\lambda}{4}\left(\tilde{\Phi}^2 - \frac{\tilde{m}^2}{\lambda}\right)^2.$$

In a very early stage of the Universe, this scalar field was in local thermodynamic equilibrium with its surroundings see [21]. At some time, the scalar field decoupled from the rest of the matter and started a lonely journey with its temperature T decreased by the expansion of the Universe. Thus, it is considered the scalar field in a thermal bath of temperature T, whose scalar field potential, extended to one loop corrections, is given by

$$V(\tilde{\Phi}) = -\frac{1}{2}\tilde{m}^2\tilde{\Phi}^2 + \frac{\lambda}{4}\tilde{\Phi}^4 + \frac{\lambda}{8}T^2\tilde{\Phi}^2 - \frac{\pi}{90}T^4 + \frac{\tilde{m}^4}{4\lambda},$$

where $\tilde{m}$ is a mass parameter before the breaking of symmetry and λ is the self-interacting constant. From here, it can be calculated the critical temperature T_c at which the Z_2 symmetry of the real SF breaks. To do that, they calculated the critical points of the scalar potential from

$$0 = \left(-\tilde{m}^2 + \lambda\tilde{\Phi}^2 + \frac{\lambda}{4}T^2\right)\tilde{\Phi}.$$

The negative term $-\tilde{m}^2$ permits the breaking of symmetry of the potential. One critical point is at $\tilde{\Phi} = 0$. If the temperature T is high enough, the scalar potential has a minimum at this critical point. Furthermore, the critical temperature T_c in which $\tilde{\Phi} = 0$ becomes a maximum is

$$T_c^2 = \frac{4\tilde{m}^2}{\lambda}.$$

This critical temperature defines the symmetry breaking scale of the scalar field.

To study the dynamics of the SFDM in the background Universe it is assumed a Friedmann–Lemaitre–Robertson–Walker metric with scale factor $a(t)$. The background Universe was composed by SFDM (Φ_0) endowed with a scalar potential $V \equiv V(\Phi_0)$, baryons (b), radiation (z), neutrinos (ν), and a cosmological constant

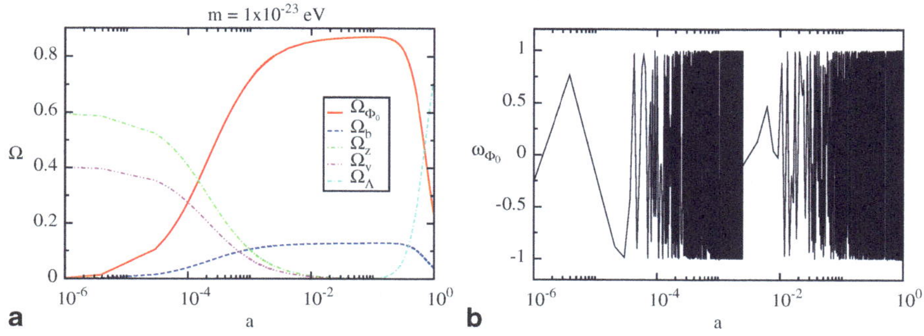

Fig. 9.1 *Left*: evolution of the density parameters Ω_i for the background universe. Scalar field dark matter model mimics the standard ΛCDM behavior. *Right*: evolution of the scalar field dark matter equation of state for the background universe

(Λ) as dark energy. For the basic background equations, we have from the energy-momentum tensor $\boldsymbol{T}$ for a scalar field, the scalar energy density T_0^0 and the scalar pressure T_j^i are given by

$$T_0^0 = -\rho_{\Phi_0} = -\left(\frac{1}{2}\dot{\Phi}_0^2 + V\right), \tag{9.1}$$

$$T_j^i = P_{\Phi_0} = \left(\frac{1}{2}\dot{\Phi}_0^2 - V\right)\delta_j^i, \tag{9.2}$$

where the dots stand for the derivative with respect to the cosmological time and δ_j^i is the Kronecker delta. Thus, the cosmological Equation of State for the scalar field is $P_{\Phi_0} = \omega_{\Phi_0}\rho_{\Phi_0}$ with (Fig. 9.1)

$$\omega_{\Phi_0} = \frac{\frac{1}{2}\dot{\Phi}_0^2 - V}{\frac{1}{2}\dot{\Phi}_0^2 + V}.$$

In order to solve the Friedmann equations with analytic methods with the approximation $m \gg H$ they performed a transformation and compared their result with numerical ones. Here the scalar field and the variables of the background depend only on time, e.g., $\Phi = \Phi_0(t)$.

They computed the growth of the SFDM overdensities $\delta\rho_\Phi$ in the linear regime, in this regime, the density contrast $\delta \equiv \delta\rho_\Phi/\rho\Phi_0$ was much smaller than unity. It is believed that the Universe was almost uniform after inflation, with a very small density contrast. As the Universe expanded, the small overdensities grew until they began to collapse, leading to the formation of structure in the Universe. Thus, only small deviations in the FLRW model are considered, so that they can be treated by linear perturbation theory. After introducing the perturbed metric tensor in the FLRW background, only scalar perturbations considered. In their paper they gave the equation of energy-momentum conservation and the Einstein field equations for

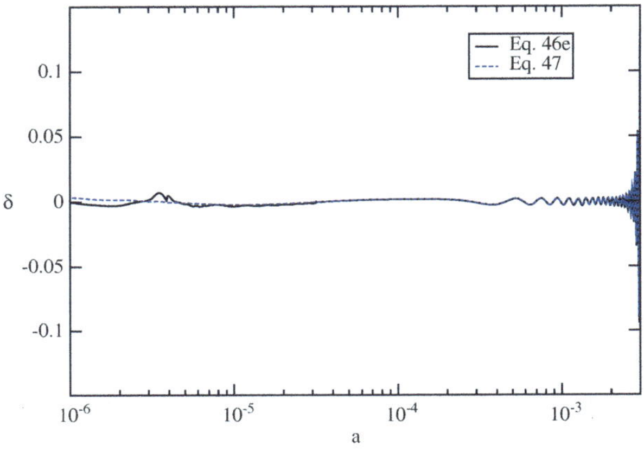

Fig. 9.2 Evolution of the density contrast δ for a perturbation with wavelength $\lambda_k \sim 2$ Mpc

the perturbed metric. Within the linear theory of scalar perturbations the evolution of the density contrast can be written as

$$\dot{\delta} + 3H\left(\left\langle\frac{\delta P_{\Phi}}{\delta\rho_{\Phi}}\right\rangle - \langle w_{\Phi_0}\rangle\right)\delta = 3\dot{\phi}_k\langle F_{\Phi}\rangle - \langle G_{\Phi}\rangle$$

where

$$F_{\Phi} = 1 + w_{\Phi_0}$$
$$G_{\Phi} = \frac{2k^2}{a^2k^2}\frac{\dot{\phi}_k + H\phi_k}{\rho_{\phi_0}}, \tag{9.3}$$

being ϕ_k the gravitational potential. This equation differs from the density contrast equation for ΛCDM. However, in [78] they show that the extra terms F_{Φ} and G_{Φ} tend to the values of the standard equation of ΛCDM. Therefore the scalar perturbations in this model grow up exactly as in the ΛCDM paradigm (Fig. 9.2).

The evolution of the scalar perturbations in the nonlinear regime when $\delta \gg 1$ was also studied. Here, an analysis was made within the framework of the spherical collapse model [94]. This formalism is very useful to understand the structure formation process in the Universe in the nonlinear regime. Focused on the era where the radiation density is equal to the SFDM density, the $T \ll T_c$, and therefore,it is expected that the scalar potential reaches the Φ_0^2 profile. They also studied if Φ^2-dark matter perturbations (once the breaking of symmetry was achieved and the SF had reached its minimum) where able to form bound structures as in the standard model.

Following the same path, [119] obtained that for the matter dominated era the low-k modes grow. When CDM decouples from radiation in a time just before recombination it grows in a milder way than it does in the matter dominated era (Fig. 9.3).

Although in general a scalar field is not a fluid, it can be treated as if it behaved like one and the evolution of its density can be the appropriate for the purpose of

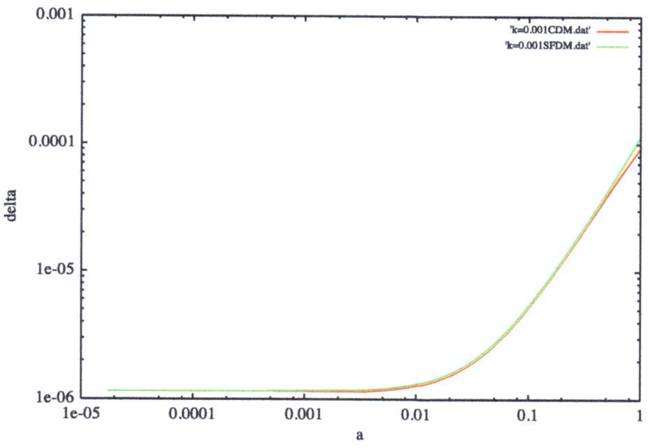

Fig. 9.3 Evolution of the perturbations for the CDM model (*red dots*) and SFDM model (green line) for $k = 1 \times 10^{-3} hMpc^{-1}$. Notice how after the epoch of equality ($a_{eq} \sim 10^{-4}$) the evolution of both perturbations is identical, $a = 1$ today. In this case the self-interacting parameter is $\lambda = 0$ [119]

structure formation, because locations with a high density of dark matter can support the formation of galactic structure.

In [119] they assumed that there was only one component to the mass density, and that this component was given by the scalar field dark matter. In this case the equation for the perturbations reads

$$\frac{d^2\delta}{dt^2} + 2H\frac{d\delta}{dt} + \left[(v_q^2 + w\hat{\rho}_0)\frac{k^2}{a^2} - 4\pi G\hat{\rho}_0\right]\delta = 0,$$

valid for all sub-horizon sized perturbations in the non-relativistic regime.

It was shown that the scalar field with an ultralight mass of 10^{-22} eV simulates the behavior of CDM in a Universe dominated by matter when $\lambda = 0$, because in general in a matter dominated Universe for low-k, v_q (called the quantum velocity) tends to be a very small quantity tending to zero, so from the equation of the density contrast we could see that on this era we have the ΛCDM profile given by

$$\frac{d^2\delta}{dt^2} + 2H\frac{d\delta}{dt} + \left(c_s^2\frac{k^2}{a^2} - 4\pi G\hat{\rho}_0\right)\delta = 0,$$

i.e., the SFDM density contrast profile is very similar to that of the ΛCDM model, Fig. 9.3. On the contrary for $\lambda \neq 0$ both models have different behavior as can be seen from Fig. 9.4, results which show that linear fluctuations on the SFDM can grow, even at early times when the large-scale modes (small k) have entered the horizon just after $a_{eq} \sim 10^{-4}$, when it has decoupled from radiation, so the amplitudes of the density contrast start to grow faster than those for CDM around $a \sim 10^{-2}$. Here an important point is that although CDM can grow it does so in a hierarchical way, while from Fig. 9.4 we can see that SFDM can have bigger fluctuations just before the ΛCDM model does, i.e., it might be that no hierarchical model of structure formation is needed for SFDM, and it is expected that for the non-linear fluctuations the behavior will be quite the same as soon as the scalar field condensates, which could be in a very early epoch when the energy of the Universe was about ~TeV. These facts can be the crucial difference between both models.

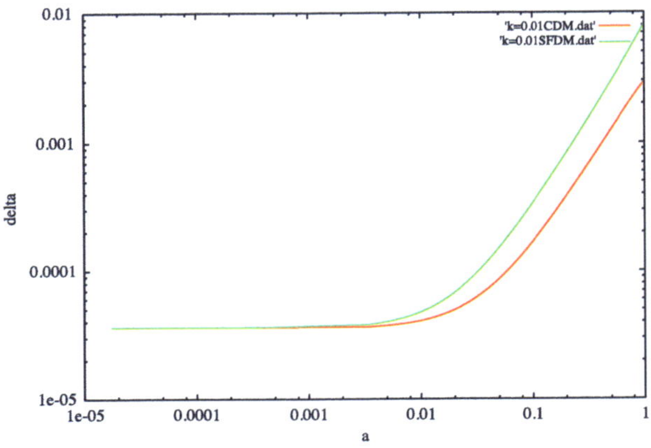

Fig. 9.4 Evolution of the perturbations for the CDM model (*red dots*) and SFDM model (*green line*) for $k = 1 \times 10^{-2} hMpc^{-1}$ and $\lambda \neq$ and negative. Notice how after the epoch of equality ($a_{eq} \sim 10^{-4}$) the evolution of both perturbations is now different from the one in Fig. 9.3, $a = 1$ today. In this case we can clearly see that the SFDM fluctuations grow faster than those for the CDM model

Additionally, in [61] the growth of cosmological perturbations to the energy density of dark matter during matter domination was considered when dark matter is a scalar field that has undergone Bose-Einstein condensation. In this case, the inhomogeneities were considered within the framework of both Newtonian gravity, where the calculation and results resulted more transparent, and General Relativity. The direction taken was again in deriving analytical expressions, which where obtained in the small pressure limit. Throughout their work the results where again compared to those of the standard cosmology, where dark matter is assumed pressureless, using analytical expressions to showcase precise differences. They also find, compared to the standard cosmology, that Bose-Einstein condensate dark matter leads to a scale factor, gravitational potential and density contrast that again increases at a faster rate.

9.4.3 Galaxies and BEC Dark Matter Mass Constraints

Magaña et al. [75] considered a model where ultra-light bosons are the main components of the dark halos of galaxies. The main goal of this work was to constrain the mass of the scalar particles. They constructed stable equilibrium configurations of SFDM in the Newtonian limit to model the DM halo in UMi. They studied two relevant cases of SFDM halos: with and without self-interaction.

Since galactic halos are well described as Newtonian systems, the work was done within the Newtonian limit. In this limit, The Einstein–Klein–Gordon equations for a complex scalar field Φ minimally coupled to gravity and endowed with a SF potential $V(\Phi) = m_\phi^2 \Phi^2/2 + \lambda \Phi^4/4$, can be simplified to the Schrödinger–Poisson equations:

$$i\hbar \partial_t \psi = -\frac{\hbar^2}{2m_\phi} \nabla^2 \psi + U m_\phi \psi + \frac{\lambda}{2m_\phi} |\psi|^2 \psi, \tag{9.4}$$

$$\nabla^2 U = 4\pi G m_\phi^2 \psi \psi^*, \tag{9.5}$$

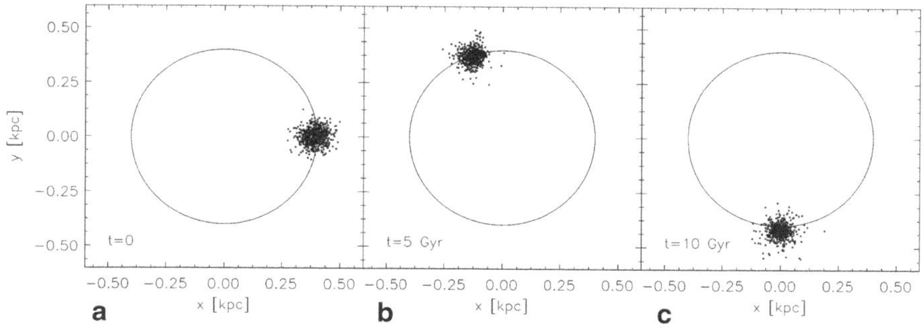

Fig. 9.5 Snapshots of the clump in UMi galaxy, at $t = 0, 5$, and 10 Gyr. The clump is set on a circular orbit in the (x, y)-plane at a distance of $= 0.39$ kpc from UMi's center. The mass of the boson is $m_\phi = 10^{-23}$ eV and $\Lambda = 0$. The total mass of the galaxy is $M = 9.7 \times 10^9 M_\odot$

where m_ϕ is the mass of the boson associated with the scalar field, U is the gravitational potential produced by the DM density core, λ is the self-interacting coupling constant, and the field ψ is related to the relativistic field Φ through

$$\Phi = e^{-im_\phi c^2 t/\hbar}\psi.$$

UMi is a diffuse dSph galaxy located at a distance of 69 ± 4 kpc from the Milky Way center and has a luminosity of $L_V = 3 \times 10^5 L_\odot$. Its stellar population is very old with an age of 10–12 Gyr. Dynamical studies suggest that UMi is a galaxy dominated by DM, with a mass-to-light ratio larger than $60 M_\odot/L_\odot$. Among the most puzzling observed properties of UMi is that it hosts a stellar clump, which is believed to be a dynamical fossil that survived because the underlying DM gravitational potential is close to harmonic. This condition is accomplished if the DM halo has a large core.

In that work, it is mentioned that the most remarkable feature in UMi structure is the double off-centered density peak. The second peak or clump is located on the north-eastern side of the major axis of UMi at a distance of ~0.4 kpc from UMi's center. The velocity distribution of the stars contained in the clump is well fitted by two Gaussians, one representing the background. The most appealing interpretation is that UMi's clump is a disrupted cluster with an orbit in the plane of the sky, which has survived in phase-space because the underlying gravitational potential is harmonic, implying that the dark halo in UMi has a large core (Fig. 9.5).

The fact that SFDM halos have cores might solve other apparent problems in dwarf galaxies. In their work the timing problem of the orbit decay of GCs (Globular clusters) in dwarf elliptical galaxies and dSph galaxies was considered. In fact, in a cuspy halo, GCs in these galaxies would have suffered a rapid orbital decay to the center due to dynamical friction in one Hubble time, forming a nucleated dwarf galaxy. For instance, under the assumption that mass follows light or assuming a NFW profile, Fornax GCs 3 and 4, which are at distances to the center < 0.6 kpc, should have decayed to the center of Fornax in $\sim$ 0.5-1 Gyr; this clearly represents a timing problem. Assuming a cuspy NFW halo, GCs 1, 2, 3 and 5 can remain in orbit as long as their starting distances from Fornax center are $\gtrsim 1.6$ kpc, whereas one of them needs an initial distance $\gtrsim$ 1.2 kpc. However, there is no statistical evidence to suggest that the initial distribution of GCs is so different to the stellar background distribution. In addition, studies of the radial distribution of GCs in giant elliptical galaxies show that the distribution of metal-rich GCs matches the galaxy light distribution. Assuming that GCs formed along with the bulk of the field star population in dwarf galaxies, the probability that Fornax GCs were formed all beyond 1.2 kpc is $\sim (0.03)^5 \simeq 2.5 \times 10^{-8}$. Therefore, it is very unlikely that all the GCs in Fornax were formed at such large distances and even if they did, there is still a timing problem.

The persistence of cold substructures in UMi places upper limits on m_ϕ. Using N-body simulations, it was found that the survival of cold substructures in UMi was only possible if $m_\phi < 3 \times 10^{-22}$ eV in the $\Lambda = 0$ case. On the other hand, by imposing a plausible upper limit on M, lower limits on m_ϕ where placed. All together, it was found that for $\Lambda = 0$, m_ϕ should be in the window

$$0.3 \times 10^{-22}\ \text{eV} < m_\phi < 3 \times 10^{-22}\ \text{eV}.$$

Since the timing problem of the orbital decay of the GCs in Fornax can be alleviated if $m_\phi < 1 \times 10^{-22}$ eV for $\Lambda = 0$, the most favored value resulted around $(0.3 - 1) \times 10^{-22}$ eV.

For SFDM models with self-interaction, the upper limit on m_ϕ increases with Λ. Bosons of mass $\lesssim 6 \times 10^{-22}$ eV could account for the observed internal dynamics of UMi. In the limit $\Lambda \gg 1$, it was found that $m_\phi^4/\lambda \lesssim 0.55 \times 10^3$ eV^4 would explain the longevity of UMi's clump and the surviving problem of GCs in Fornax.

The window of permitted values for m_ϕ resulted quite narrow. Even so, it is remarkable that the preferred range for the mass of the boson derived from the dynamics of dSph galaxies resulted compatible with those given by other authors to ameliorate the problem of overabundance of substructure and is also consistent with the CMB radiation [56, 80, 110].

In a recent posting, Slepian and Goodman constrained the mass of bosonic DM using rotation curves of galaxies, and Bullet Cluster measurements of the scattering cross section of self-interacting DM under the assumption that these systems are in thermodynamic equilibrium, [116]. If their assumptions are verified, repulsive bosonic DM will be excluded and, thereby, the only remaining window open is non-interacting bosons. Nevertheless, the static diffusive equilibrium between Bose-Einstein condensate and its non-condensated envelope, as well as finite temperature effects need to be reconsidered. In addition, other authors [106, 111] argue that scattering cross sections for bosonic DM are much smaller than those derived from the condition of thermodynamic equilibrium by Slepian and Goodman.

9.4.4 SFDM Halos and Lensing

There are two observations of galaxies that can give us some information about the nature of dark matter; rotation curves and gravitational lensing,these two related with the presence of dark matter. While the first one can be studied using the Newtonian limit, the second one results completely relativistic. Each one separately can not determine the nature of DM, but together can give us important information about this open problem.

In their work, [91], Núnez et al. used a static and spherically symmetric metric to model DM halos. The metric had two free functions, one associated with the distribution of mass and the other one with the gravitational potential. They used galactic, typical rotation curves to determine the kinematics of the halos. They calculated the mass functions for a perfect fluid and a scalar field, separately and demonstrated that both models can fit the observations. They then employed lensing to discriminate between the models.

With these examples it was shown how a perfect fluid and a scalar field can be consistent with the observations of rotation curves of DM halos, though they lead to different conclusions to the mass function. The deflection of light was then used to discriminate between the two models. Even though the mass function for some models did not have the intuitively expected behavior, it was necessary to use the observations in order to discard the model.

Recently Gonzále-Morales et al. presented an observational constraint to the model of SFDM arising from strong lensing observations in galaxies, [40]. Their result pointed to a discrepancy in the properties of SFDM halos for dwarf and lens galaxies, mainly because halo parameters resulted related to the physical quantities of the model.

They showed that a discrepancy between lensing and dynamical studies appeared if they considered that the SFDM mass density profile in

$$\rho(r) = \begin{cases} \rho_c \frac{\sin(\pi r/r_{max})}{(\pi r/r_{max})} & \text{for } r < r_{max} \\ 0 & \text{for } r \geq r_{\max} \end{cases}$$

described the inner regions of galactic halos at different redshifts, up to radii of order 5–10 Kpc. They found that lensed galaxies at $z \sim 0.5$, if described by a SFDM halo profile, should be denser than dwarf spheroidals in the local universe, in order to satisfy the conditions necessary to produce strong lensing.

On the other hand more recently, Robles and Matos have investigated the gravitational constraints imposed to dark matter halos in the context of finite temperature scalar field dark matter, [109]. They gave a strong lensing constraint of a finite temperature scalar field DM halo.

They showed that there are differences with respect to the full Bose–Einstein condensate halo when the temperature of the scalar field in dark matter halos is taken into account. They extended the previous analysis of a fully condensed system at temperature zero, and showed that multiples images are possible with more than one state of the scalar field, i.e., the non-zero temperature allows the scalar field to be in excited states. As finite temperature DM halos are not of only one radius, then, their constraint expresses two limits, either the halos of strong lensing systems are 10 times larger or 10 times denser than dwarf galaxy halos.

They also provided a way to identify the excited state of the DM halo by means of measuring their Einstein radius, the closer it is to the center the more probable that the SFDM halo is in a higher excited state. A deeper analysis can be used as a test to the validity of the SFDM model, mainly, because identifying the excited states of various halos can give information about their evolution which can be compared with simulations and work as test to the model.

9.4.5 SFDM Density Profiles and LSB Galaxies

An analysis of the Newtonian regime at Temperature zero can be found in Bohmer and Harko (2007), they assumed the Thomas–Fermi approximation which neglects the anisotropic pressure terms that are relevant only in the boundary of the condensate, the system of equations describing the static BEC in a gravitational potential V is given by

$$\nabla p \left(\frac{p}{m}\right) = -p\nabla V, \tag{9.6}$$

$$\nabla^2 V = 4\pi G\rho, \tag{9.7}$$

with equation of state

$$p(\rho) = U_0 \rho^2,$$

where $U_0 = \frac{2\pi\hbar^2 a}{m}$, ρ is the mass density of the static BEC configuration and p is the pressure at zero temperature, p is not the usual thermal pressure but instead it is produced by the strong repulsive interaction between the ground state bosons.

In [107], Robles and Matos found that the BEC dark matter model can give a density contrast profile consistent with RC's of dark matter dominated galaxies. The profile resulted as good as one of the most frequently used empirical core profiles, the pseudo Isothermal profile (PI), but with the advantage of coming from a solid theoretical frame. In [107] The data was fitted within 1 kpc and a logarithmic slope $\alpha = -0.27 \pm 0.18$ was found in perfect agreement with a core. They emphasized that the cusp in the central regions is not a prediction that comes from first principles in the CDM model, it is a property that is derived by fitting simulations that use only DM. For a detail discussion of the cusp and core problem in the standard model see [32] and references there in. They also explained an ambiguity in the usual interpretation of the core radius, they proposed a new definition for the core and core radius that takes away this ambiguity and that has a clear meaning that allows for a definite distinction of when a density profile is core or cusp. Using their definition they found the core radius in the BEC profile to be in most cases over 2 kpc bigger than the core radius in the PI profile. They assumed a great number particles were in the ground state in the form of a condensate. This led to good results for their sample of galaxies, but it proved necessary to consider more than these simple hypotheses when dealing with large galaxies.

The solution to the system above is

$$\rho_B(r) = \rho_0^B \frac{sin(kr)}{kr} \tag{9.8}$$

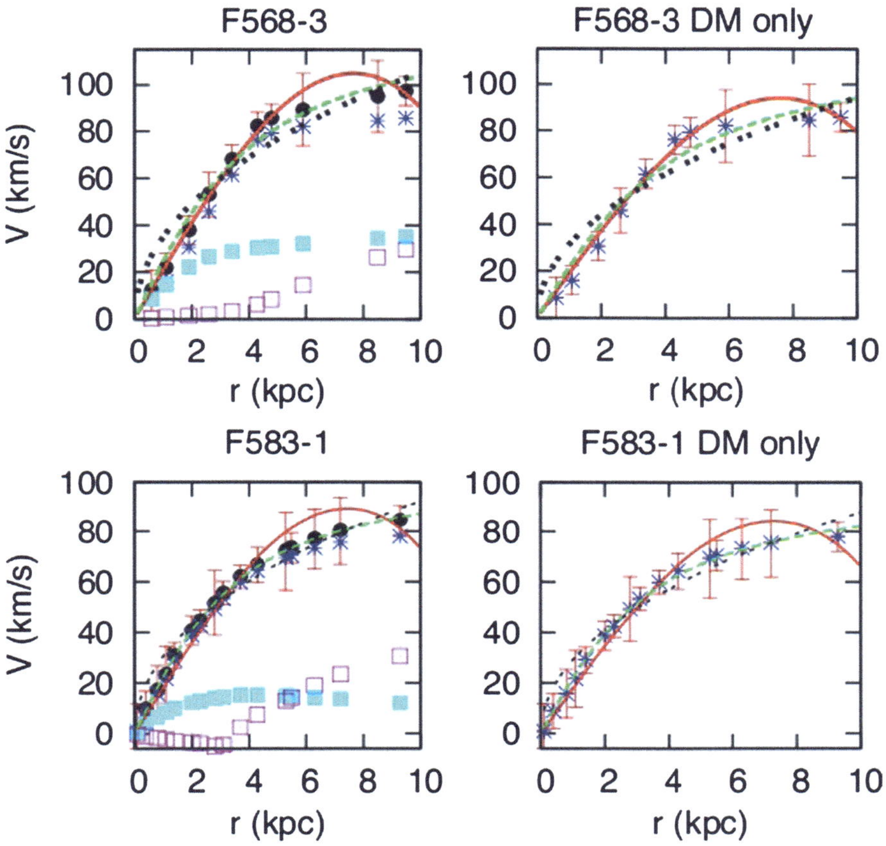

Fig. 9.6 Contribution of the baryons to the rotation curve for F568-3 and F583-1. We denote observed data by *black dots* with error bars, dark matter with blue asterisks, the disk with cyan squares and the the gas with magenta squared boxes. The figures on the left were fitted asumming the minimum disk hypothesis while the ones on the right are only the dark matter. In the fits shown are, BEC in *solid line*(*red* in the online version), PI dashed (*green* in the online version) and NFW double-dashed (*black* in the online version) profiles

it can be seen that the BEC model satisfies $\rho \sim r^0$ near the origin, but a priori this does not imply consistency with observed RC's. Therefore, the profiles were fitted to thirteen high resolution RC data of a sample of LSB galaxies. The RC's were taken from a subsample of de Blok et al. (2001), galaxies that have at least 3 values within $\sim$1 kpc where chosen, not presenting bulbs and the quality in the RC in H_α is good as defined in McGaugh S. S. et al. (2001). Because the DM is the dominant mass component for these galaxies they adopt the minimum disk hypothesis which neglects baryon contribution to the observed RC. In order to show that in LSB and dwarf galaxies neglecting the effect of baryons was a good hypothesis, they included two representative examples, see Fig. 9.6.

For these galaxies the contribution of the gas was plotted, disc and dark matter separately. They did the fitting first considering the total contribution and then using only DM. They found no substantial difference in their values.

As a second result and direct consequence of the core definition, they where able to obtain the constant value of μ_0 which is proportional to the central surface density. This result is one of the conflicts of the current standard cosmological model due to the hierarchical formation of galaxies.

As the density profile Eq. (9.8) is not enogh to describe the large galaxies as discussed in [108], Robles and Matos thus gave a physically motivated extension to the SFDM model that includes the DM temperature corrections to the first loop in perturbations. Their idea is to use the Z_2 spontaneous symmetry break of a real scalar field as a new mechanism in which the early DM halos form. As stated earlier, when the real scalar field rolls down to the minimum of the potential, the perturbations of the field can form and grow. They gave an exact analytic solution for an static spherically symmetric SF configuration, which in the SFDM model represents a DM halo. Their solution naturally presents a flat central density profile,just as Eq. (9.8), but now it can accommodate more than just the ground state as the temperature $T \neq 0$, in this way they solved previous discrepancies in rotation curve fits at $T = 0$, for instance, having a constant halo radius for all galaxies and the incapability to fit at the same time the inner and outermost regions of RC in large galaxies. Both issues were solved using this scenario which includes temperature of the DM and the exited states of the SF.

The perturbed system of a scalar field with a quartic repulsive interaction but with temperature zero has been studied before [28, 38]. The study for the evolution of the SF with the temperature correction in a FRW universe is analogous. The metric tensor was written as $\boldsymbol{g} = \boldsymbol{g}_0 + \delta \boldsymbol{g}$, where as always $\boldsymbol{g}_0$ is the unperturbed FRW background metric and $\delta \boldsymbol{g}$ the perturbation. The perturbed line element in conformal time η given by

$$ds^2 = a(\eta)^2[-(1+2\psi)d\eta^2 + 2B_{,i}\, d\eta dx^i + (1-2\phi)\delta_{ij} + 2E_{,ij}\, dx^i dx^j]$$

with a the scale factor, ψ the lapse function, ϕ gravitational potential, B the shift, and E the anisotropic potential. The energy-momemtum tensor and the field where separated as $\boldsymbol{T} = \boldsymbol{T}_0 + \delta \boldsymbol{T}$ and $\Phi(x^\mu) = \Phi_0(\eta) + \delta\Phi(x^\mu)$ respectively. As the linear regime was studied $\delta\Phi(x^\mu) \ll \Phi_0(\eta)$, the approximation $V(\Phi) \approx V(\Phi_0)$ could be made. They worked in the Newtonian gauge where the metric tensor $\boldsymbol{g}$ becomes diagonal and as a result, in the trace of the Einstein's equations the scalar potentials ψ and ϕ are identical, therefore, ψ relates to the gravitational potential.

The work was mainly focused in the galactic scale DM halos after their formation. Robles and Matos constrained themselves to solve the Newtonian limit of the perturbed KG equation, that is,

$$\Box\delta\Phi + \frac{\hat{\lambda}}{4}[k_B^2(T^2 - T_C^2) + 12\Phi_0^2]\delta\Phi - 4\dot{\Phi}_0\dot{\phi} + \frac{\hat{\lambda}}{2}[k_B^2(T^2 - T_C^2) + 4\Phi_0^2]\Phi_0\phi = 0$$

valid when Φ is near the minimum of the potential and after the SB, where the SF is expected to be stable. Here T_C is the temperature of the symmetry break.

Fig. 9.7 Rotation curve of NGC 1560 and NGC1003 from [108], where non zero temperature is considered. The wiggles are visible in the outer region of the RCs. The solid line is the temperature corrected velocity profile and the dashed line is the Einasto fit [108]

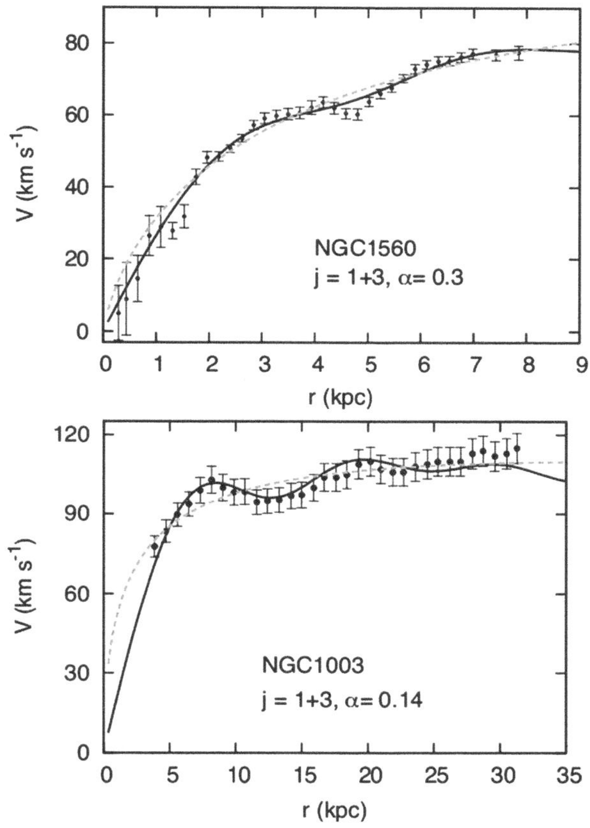

Additionally to solving these two disagreements they mentioned why it does not seem necessary to include high amounts of feedback to fit and reproduce the inner core and wiggles found in high-resolution RC's, see Fig. 9.7. Also, this model can be tested with high redshift observations, the SFDM model predicts initial core profiles as opposed to the initially cuspy ones found in CDM simulatios which are expected to flatten due to redistribution of DM by astrophysical processes.

Finally one conclusion is that if observations of more galaxies with core behavior are confirmed, this model can be a good alternative to ΛCDM.

9.4.6 BEC Dark Matter and the Power Spectrum

From another point of view in [125] it was again assumed that dark matter is composed of scalar particles that are able to form a Bose-Einstein condensate at some critical redshift z_{cr}, but in this case it was used to study the matter power spectrum.

After the BEC forms its effective pressure can assume a polytropic equation of state such as $P_{be} \sim \rho_{be}^{\gamma}$ if an arbitrary non-linearity term is assumed, in this case γ.

The exact value of γ is defined by the non-linear contribution of the Gross-Pitaevskii equation which in its standard form leads to [18]

$$\rho_{be} = \frac{2\pi\hbar^2 l_a}{m_\chi^3}\rho_{be}^2.$$

In this case, the scattering length l_a and the mass m_χ of the dark matter particle again determine the dynamics of the fluid. Assuming that the condensate does not interact with any other form of energy, the above pressure, via the conservation balance, leads to

$$\rho_{be} = \frac{m_\chi^3}{2\pi\hbar^2 l_a}\frac{\rho_0}{a^3 - \rho_0}$$

where

$$\rho_0 = \frac{1.266 \times \Omega_{be0} \times (m_\chi/1\ \mathrm{meV})^{-3} \times (l_a/10^9\ \mathrm{fm})}{1 + 1.266 \times \Omega_{be0} \times (m_\chi/1\ \mathrm{meV})^{-3} \times (l_a/10^9\ \mathrm{fm})}.$$

The current value of the scale factor a was taken as $a_0 = 1$ and the current fractional density of the BEC dark matter as $\Omega_{be0} = \rho_{be0}/\rho_c$ where ρ_c is the critical density. When the above relations are combined the equation of state parameter of the BEC dark matter is obtained

$$w_{be} = \frac{\rho_0}{a^3 - \rho_0}.$$

If the inertial effects of the pressure become relevant as for example during the radiation phase or at the onset of the accelerated expansion, Newtonian cosmology fails and a more appropriate set of equations is needed. The inclusion of pressure in the Newtonian cosmology gives rise to the neo-Newtonian cosmology. In this case, the matter power spectrum is defined as always

$$P(k) = |\delta_b(z = 0; k)|^2,$$

where $\delta_b(k)$ is the baryonic density contrast calculated from equations

$$\delta_b'' + \delta_b'\left(\frac{H'}{H} + \frac{3}{a}\right) - \frac{3}{2}\frac{\Omega_b}{H^2a^2}\delta_b = \frac{3\Omega_{be}}{2\,H^2a^2}(1 + c_s^2)\delta_{be},$$

and

$$\delta_{be}'' + \left(\frac{H'}{H} + \frac{3}{a} - \frac{w_{be}'}{1 + w_{be}} - \frac{3w_{be}}{a}\right)\delta_{be}' + \left[3w_{be}\left[\frac{H'}{Ha} + \frac{(2 - 3w_{be})}{a^2}\right]\right.$$
$$\left. + \frac{3w_{be}'}{a(1 + w_{be}) + \frac{(k/k_0)^2 c_s^2}{H^2a^4}} - \frac{3}{2}\frac{\Omega_{be}}{H^2a^2}(1 + 3c_s^2)(1 + w_{be})\right]\delta_{be}$$
$$= \frac{3}{2}\frac{\Omega_b}{H^2a^2}(1 + c_s^2)\delta_b \tag{9.9}$$

at the present time [125]. The baryonic agglomeration δ_b is supposed to be driven by the gravitational field which is sourced by all the forms of energy. In order to solve this set of equations it was needed to set the initial conditions for δ_b and δ_{be} and their derivatives at z_{cr} where the condensation took place. Since $z_{cr} = z_{cr}(l_a, m_\chi)$ for each chosen couple of values (l_a, m_χ) different initial conditions where needed.

In [125] it was shown that if such phase transition occurred in the recent Universe this process would be able to leave small, but perceptible, imprints on the large scale structure perceptible in the matter power spectrum. Assuming $l_a = 10^6$ fm the BEC dark matter model does shows differences of the order of a few percents for masses 15–35 meV. Adopting $l_a = 10^{10}$ fm corrections of the same order where obtained for masses 300–700 meV.

Although the BEC phase is shown to have a small influence on the matter power spectrum, a more quantitative analysis could be performed to estimate the preferred values of the model parameters.

For the relevant parameter values studied in that work, the transition to the BEC phase was shown to occur at low redshifts. Since the standard cosmology remains unchanged before z_{cr} the CMB physics at the last scattering surface remained the same. However, the BEC dark matter would modify the gravitational potential just after z_{cr} while the speed of sound is nonzero leading to a contribution to the integrated Sachs–Wolfe effect.

9.4.7 Vortices in BEC Dark Matter

The conditions to form vortices in a SFDM/BEC halo have also been studied in [106, 130]. As it has been pointed out, these halos can be described as fluids, obeying quantum-mechanical fluid equations, so that the effects that make up this form of dark matter behave differently from standard ΛCDM, resulting in new effects with potentially observable consequences. The idea is that ultralight particles with $m \ll 1$ eV will have very large de Broglie wave lengths which means that quantum statistical effects are important and macroscopic coherent lumps of matter can emerge. These light Bose particles will have a transition temperature to the condensed state that is of order $T_c \sim 2 \cdot 10^9$ K, which is the expected temperature in the Universe after about 1 s.

There are essentially two limiting cases that can be considered. First, for quantum-coherence to be relevant on the scale of a halo of radius R, the particle de-Broglie wavelength

$$\lambda_{deB} = \frac{h}{mv},$$

should be considered to be of the order of the halo size, $\lambda_{deB} \lesssim R$, or else require $\lambda_{deB} \ll R$ but with a strong repulsive self-interaction to hold the halo up against gravity. In the first case, if $v \simeq v_{virial}$ for the halo, this translates into a condition for the dark matter particle mass,

$$m \gtrsim m_H = 1.066 \cdot 10^{-25} (R/100\ \text{kpc})^{-1/2} (M/10^{12} M_\odot)^{-1/2}\ \text{eV cm}^3,$$

where m_H is a mass that depends on the properties of the halo and g_H is determined by the density and the radius of the halo. In the second case, for the repulsive self-interaction pressure force to exceed the quantum pressure, it is required that $g \gg g_H = 2.252 \cdot 10^{-64} (R/100\ \text{kpc})(M/10^{12} M_\odot)^{-1}\ eV cm^3$. If R is taken to be the radius of the virialized object supported against gravity by the dominant repulsive self-interaction, this imposes a condition on the particle mass given by $m \gtrsim \frac{m_H}{4}\sqrt{15\, g/g_H}$.

However, it seems that rotating BEC haloes add new phenomenology, and the possibility to distinguish this form of dark matter from other candidates. To this aim, in [106] they have studied the question of whether an angular velocity could be sufficient to create vortices in BEC/CDM cosmologies. As quantum fluid systems, BEC haloes can be modeled as uniformly rotating ellipsoids, with and without internal motions superposed. To this aim, in [106] the authors derived equations which relate the eccentricities of haloes to their λ-spin parameter. Once the latter is fixed, the eccentricities can be uniquely determined. They analytically studied necessary and sufficient conditions for vortex formation. In their results they found that vortex formation requires as a necessary condition that the halo angular momentum satisfies $L \geq L_{QM} = N\hbar$, which implies a lower bound on m/m_H, i.e. on the dark matter particle mass. However, a sufficient condition for vortex formation could be established by an energy analysis, which aimed to find the conditions of when a vortex becomes energetically favored.

They studied two classes of models for rotating halos in order to analyze stability with respect to vortex formation in two limits, one for $L/L_{QM} \gg 1$ and for $L/L_{QM} = 1$, respectively. In what they called Halo-Model A ($L/L_{QM} \gg 1$) these where modeled as homogeneous Maclaurin spheroids. The minimum angular momenta for vortex formation in this case was $(L/L_{QM})_{crit} = (5.65, 4.53, 4.02)$ for $\lambda = (0.01, 0.05, 0.1)$, respectively, which corresponded to a constraint on the particle mass $m/m_H \geq (m/m_H)_{crit}$, where $(m/m_H)_{crit} = (309.41, 49.52, 21.73)$, respectively. As long as m/m_H satisfied this condition, the strength of the self-interaction also satisfied the condition $g/g_H \geq (g/g_H)_{crit}$, where $(g/g_H)_{crit} = (1.02 \cdot 10^5, 2549.24, 454.54)$ for the same λ-values, respectively.

For Halo-Model B ($L/L_{QM} = 1$), which was an ($n = 1$)-polytropic Riemann-S ellipsoid, strictly irrotational prior to vortex formation, even $L/L_{QM} = 1$ resulted to be sufficient for vortex formation if the self-interaction strength was large enough. The condition $L/L_{QM} = 1$ fixed the value of m/m_H for each λ according to

$$\begin{aligned}\frac{L}{L_{QM}} &= \frac{m}{m_H}\frac{\kappa_n}{10}\frac{2\tilde{\Omega}\sqrt{1-e_1^2 e_1^4}}{(2-e_1^2)(1-e_1^2)^{5/6}(1-e_2^2)^{1/3}} \\ &= \frac{m}{m_H}\frac{\kappa_n}{10} \times \left(\frac{2B_{12}}{q_n}\right)^{1/2}\left(2+\frac{e_1^4}{4(1-e_1^2)}\right)^{-1/2}\frac{e_1^4}{(1-e_1^2)^{5/6}(1-e_2^2)^{1/3}},\end{aligned} \tag{9.10}$$

and the condition of virial equilibrium

$$y(x) = \frac{\pi}{\sqrt{8}} g(e_1, e_2)^{-1/2} x,$$

thereby also fixing g/g_H. For $\lambda = (0.01, 0.05, 0.1)$, these values where given by $m/m_H = (44.58, 9.49, 5.01)$ and $g/g_H = (1595.07, 68.00, 17.20)$, respectively. Halo-Model B then made vortex formation energetically favorable for these values of m/m_H and g/g_H. They interpreted this to mean that, for $L/L_{QM} > 1$, vortex formation will also be favored, as long as $g/g_H > (g/g_H)$. Furthermore, any values of m/m_H and g/g_H which satisfy the condition for vortex formation in Halo-Model A would automatically satisfy that found by Halo-Model B, which resulted less stringent but more accurate.

In conclusion they imagined vortex formation in BEC haloes composed of repulsively interacting particles as follows: If the angular momentum of a rotating BEC halo fulfills $L < L_{QM}$, no vortex would form, and the halo can be modeled by a mildly compressible, irrotational Riemann-S ellipsoid, which has a polytropic index of $n = 1$. For $L = L_{QM}$, the irrotational Riemann-S ellipsoidal halo can make a transition to a non-rotating, spherical halo with a vortex at the center if the self-interaction is strong enough. For a range of angular momenta fulfilling $L_{QM} < L \leq 2L_{QM}$, a central vortex can be expected but now with the excess angular momentum deforming the halo such that again a Riemann-S ellipsoid forms. Finally, if $L \gg L_{QM}$, oblate haloes described as Maclaurin spheroids had a central vortex if $m/m_H \geq (m/m_H)_{crit}$ and $g/g_H \geq (g/g_H)_{crit}$ with the critical values given by Halo-Model A. Those critical values determined when a single vortex was energetically favored, but since $L/L_{QM} \gg 1$, it is also possible that multiple vortices could form.

From another point of view, in [130] it was assumed that the particles where non-interacting and therefore only gravity acted on the system. Following [56], the authors based on Jeans instability analysis to estimate their parameters. The growing mode under gravity was given by $e^{\gamma t}$ with $\gamma^2 = 4\pi G\rho$, whereas the free field was supposed oscillatory; e^{-iEt} with $E = k^2/2\,m$. The latter was then written as $e^{\gamma t}$ with $\gamma^2 = -(k^2/2\,m)^2$. Noting that this is like normal Jeans analysis with sound speed $c_s^2 = k/2\,m$ then $\gamma^2 = 4\pi G\rho - (k^2/2\,m)^2$. Setting this to zero, for the Jeans scale they found

$$r_J = 2\pi/k_J = \pi^{3/4}(G\rho)^{1/4} m^{-1/2} = 55 m_{22}^{-1/2} (\rho/\rho_b)^{-1/4} (\Omega_m h^2)^{-1/4}\ \text{kpc},$$

where $m_{22} = m/10^{-22}$eV, and the background density is $\rho_b = 2.8 \cdot 10^{11} \Omega_{mh}^2 M_\odot$ Mpc^{-3}. It is supposed that below the Jeans scale the perturbations are stable and above it they behave as ordinary CDM [56]. As always, the stability below the Jeans scale was guaranteed by the uncertainty principle. If the particles are confined further, their momenta increases and oppose gravitational contraction.

Moreover, in [130] the author also considered the suggestion that superfluid BEC dark matter in rotation would likely lead to vortices as seen in atomic BEC experiments. As already noted in [41], BEC dark matter with self-interactions could actually constitute a superfluid. In the case of a repulsive self-interaction in [130] the author too argued that the vortex size should be determined locally by the coherence length. This means that there would be two scales in the problem: A galactic one, given by the de Broglie wavelength from the tiny mass, and a sub galactic one determined by the mass and the two-body interaction strength. He also explored the consequences of self-interactions on the virialization of gravitationally bound structures and found almost no effects for reasonable values of m and a. Here the case considered was also that of a quartic self-interaction $\lambda\phi^4$.

In [18] there was a brief discussion of the effect on the Lane–Emden equation, whereas in other works a BEC of axions with a single vortex arising from global rotation in the early Universe has been considered. This latter scenario is, however, less likely to occur since the global rotation rate of the Universe can be estimated from various observations and is very small but nevertheless non-zero.

Under the assumption of dark matter being an ultra-light BEC, the rotation of spiral galaxies would cause vortex lattices to form. In [130], the author also considered possible effects of sub galactic vortices in the dark matter on the rotation velocity curves of virialized galaxies with standard dark matter halo profiles. He found that one can actually get substructure in the rotation curves that resemble some observations, but that this required large vortex core size and small vortex-vortex distances. The mass and interaction strength needed to realize this were found to be fine-tuned, but could possibly be accommodated in more general setups.

If dark matter contains a component of condensed BEC particles that is superfluid and if the halos are rotating then it is not inconceivable that there can be vortex formation. However, the quantized vortex discussion makes an important assumption about the coherence length, ξ, entering Ω_c (angular velocity of the halo) in

$$\Omega_c = \frac{\hbar}{mR^2}\ln\left(\frac{1.46R}{\xi}\right) = \frac{6.21\cdot 10^{-17}}{m_{22}R^2}\ln\left(\frac{1.46R}{\xi}\right).$$

Here ξ is taken to be of kpc size. It then results that with no self-interaction there is only the gravitational scale $\hbar^2/GMm^2$ available, which becomes of galactic size for masses $m \sim 10^{-22}$eV. However, when including self-interactions through the scattering length a, there is also a scale given by $\xi = 1/\sqrt{8\pi an}$, which is the usual Gross-Pitaevskii coherence length. The latter coincided with the characteristic length over which the density is expect to go to zero in a vortex.

If the additional assumption that the vacuum expectation value, ϕ_0, arises from a mechanism that preserves parity, the interaction term results

$$\frac{(mc^2)^2\phi^4}{4(\hbar c)^2\phi_0^2} = g\frac{\phi^4}{4}.$$

This terms is of course merely the standard interaction term in the Gross–Pitaevskii theory of interacting condensed bosons. Therefore he also concludes that the self-interacting scenario emerges from this procedure.

These investigations and simple numerical experiments pointed to an interesting effect that could arise from bosonic dark matter. However, to fully explore the influence that vortex lattice formation and stellar feedback on structure formation has in luminous matter, an ultralight BEC dark matter component in large N-body simulations should be considered.

9.4.8 Bose Dark Matter Additional Constraint

In [111] BE condensation inside the primeval fireball, at zero-order in perturbation theory has been studied. Here, the process of condensation was considered to be driven by self-interactions of high-energy bosonic particles.

In this work it was found that in the instantaneous decoupling approximation, the subsequent evolution of the full bosonic system was only affected by the expansion rate of the Universe and small gravitational instabilities.

The evolution of bosonic DM after decoupling was analyzed as follows: their velocity and temperature affected only by the expansion rate of the Universe.

In

$$m = \frac{\Omega_c \rho_{cr}}{n_c^{(0)}} = \frac{\Omega_H \rho_{cr}}{n_T^{(0)}},$$

the temperature was needed in order to calculate the mass of the bosonic DM. Here Ω_c and Ω_H represent the content of bosonic CDM and bosonic HDM (hot dark matter) respectively, n_c is the number density of condensed bosons and n_T the number density of thermal bosons.

As an additional remark, they also address a bound on the strength of the bosons self-interactions.

In their study, the bosonic DM parameters where addressed as the mass, m, and the factor g_x (amount of degrees of freedom); where bounds on their values have been obtained from a statistical analysis of cosmological data. The constraints found for the temperature of the boson gas $T_0^{\phi} = 2.14 \pm 0.02$K, and for the boson-antiboson gas $T_0^{\phi\bar{\phi}} = 1.91 \pm 0.05$K.

Finally, from a similar analysis for fermionis they found bounds in the sum of neutrino masses and the number of extra relativistic species, $\sum m_\nu \lesssim 0.45$eV, $N = 1.10 \pm 0.18$, in concordance with some previous reports.

In summary, they presented a generic study of DM based on BE condensation, from which, CDM and HDM, result intrinsically related.

9.4.9 SFDM and Black Holes

The rapid decay of the energy density of the scalar field for the case of supermassive black holes, indicates that scalar fields may not be maintained around a black hole during cosmological time scales in the whole space; so either the scalar field gets accreted or it escapes through future null infinity. The fact is that when a

Schwarzschild black hole that is asymptotically flat is considered, scalar fields tend to vanish from the spatial domain.

An appropriate coordinate system to study the propagation of scalar fields in the Schwarzschild space-time is using hyperboloidal slices, because it has been seen that such slices reach future null infinity instead of spatial infinity, which results as a natural boundary for a wave-like processes, including electromagnetic fields and gravitational radiation.

In this case the Klein-Gordon equation for a scalar field $\tilde{\phi}_T$ can be written as [30]: $\Box\tilde{\phi}_T - \frac{d\tilde{V}}{d\tilde{\phi}_T} = 0$, where $\Box\tilde{\phi}_T = \frac{1}{\sqrt{-\tilde{g}}}\partial_\mu[\sqrt{-\tilde{g}}\tilde{g}^{\mu\nu}\partial_\nu\tilde{\phi}_T]$, with a potential of the form $\tilde{V} = \frac{1}{2}m_B^2|\tilde{\phi}_T|^2 + \frac{\lambda}{4}|\tilde{\phi}_T|^4$, where m_B has the units of mass. Then the KG equation in the conformal metric can be expressed as:

$$\tilde{\Box}\tilde{\phi}_T - \frac{1}{6}\tilde{R}\tilde{\phi}_T - (m_B^2\tilde{\phi}_T + \lambda\tilde{\phi}_T^3) = \Omega^3$$

$$\left[\Box\phi_T - \frac{1}{6}R\phi_T - (m_B^2\Omega^{-2}\phi_T + \lambda\phi_T^3)\right] = 0,$$

provided the relationship between the physical scalar field $\tilde{\phi}_T$ and the conformal scalar field ϕ_T to be $\phi_T = \tilde{\phi}_T/\Omega$. Here $R = \frac{12\Omega}{r^2}[r + (2r - 1)]$ is the Ricci scalar of the conformal metric and $\Box = \nabla^\mu\nabla_\mu$ corresponds to the conformal metric. The case for which this last equation results conformally invariant corresponds to the zero mass case $m_B = 0$.

In their work [30], used initial scalar field profiles with $m_B^2 = 0.0, 0.1, 0.2$. In order to explore the parameter space, various values of the amplitude $A = 0.01, 0.1, 1.0$ and different widths of a Gaussian pulse, which in physical units corresponded to $\sigma_1 = 0.5$, $\sigma_2 = 1$ and $\sigma_3 = 5$ to the right from $r_0 = 0.8$. This range of parameters allowed the authors to cover length ranges that involve Compton wave-length related to effects of interaction and reflection that may involve reflection and absorption effects.

In their work, [30], one of the conclusions was that one potential ingredient that would help at maintaining massive scalar field densities during longer times is the rotation of the black hole and also of the scalar field. It would then be of major interest to use foliations that approach future null infinity and study if the same effects occured and also the study of scalar field configurations with non-zero angular momentum. In fact, the results in their work where obtained assuming the maximum cross section of accretion due to the spherical symmetry, which in turn worked as upper bounds accretion rates in more general cases.

Yet another possibility results on considering solutions that asymptotically may contain a cosmological constant, which would be appropriate if a background energy density in the universe where assumed. This would imply that black hole candidates should not be considered to be asymptotically flat. Other possibilities may include black hole candidates of a different nature like boson stars. For further reading on subjects related to black holes with scalar fields and gravitational instability see for example, [62].

9.4.10 Threats to Scalar Field Dark Matter, Black Holes?

As mentioned in the last subsection, the existence of a long-lasting scalar field configurations surrounding a black hole have been studied [13, 30]. Another motivation for these studies is the possibility that super-massive black holes at galactic centers may represent a serious threat to the scalar field dark matter models.

As a first step, a relatively simple model has been considered [13]. There, stationary scalar field configurations to the Klein–Gordon equation

$$(\Box - \mu^2)\phi = 0,$$

where looked for on a Schwarzschild space-time background,

$$ds^2 = -N(r)dt^2 + \frac{dr^2}{N(r)} + r^2 d\Omega^2, N(r) := 1 - 2M/r,$$

with the d'Alambertian operator defined as $\Box := (1/\sqrt{-g})\partial_\mu(\sqrt{-g}g^{\mu\nu}\partial_\nu)$, M being the mass of the black hole and $d\Omega^2 := d\theta^2 + \sin^2\theta d\varphi^2$ the standard solid angle element. So far, the case has been restricted to the case of a canonical, massive, non self-interacting minimally coupled scalar field ϕ. With these conventions ϕ results dimensionless, while μ has dimensions of length^{-1}. The associated quantum mechanical "mass" of the scalar field given by $\hbar\mu$.

In order to look for the stationary solutions of

$$\left[\frac{1}{N(r)}\frac{\partial^2}{\partial t^2} - \frac{\partial}{\partial r}N(r)\frac{\partial}{\partial r} + \mathcal{U}_l(\mu, M; r)\right]\psi_{lm} = 0,$$

a further decomposition of the functions $\psi_{lm}(t, r)$ was done into oscillating modes of the form:

$$\psi_{lm}(t, r) = e^{i\omega_{lm}t}u_{lm}(r),$$

with ω_{lm} a real frequency and $u_{lm}(r)$ a complex function of r in the interval $(2M, \infty)$.

Although stationary solutions were found for the scalar field,

$$||u||^2 := \int_{2M}^{\infty}\left(N(r)\left|\frac{\partial u}{\partial r}\right|^2 + \mathcal{U}_l(\mu, M; r)|u|^2\right)dr$$

they have been shown to be unphysical, in the sense that their energy density integrates to infinity in a compact region just outside the event horizon. However, there seem to exist long-lasting, quasi-stationary solutions of finite energy, which are found by evolving initial data that was constructed by slightly modifying a particular subset of the stationary solutions. The solutions found so far show as an overall behavior an exponential energy decay, caused by scalar field leaking into the black hole, that in some cases can be very slow.

The stationary solutions where obtained by solving a time-independent Schrödinger-like equation

$$\left[-\frac{\partial^2}{\partial r^{*2}} + V_{eff}(r^*)\right] u(r^*) = \omega^2 u(r^*), \; -\infty < r^* < \infty,$$

with an effective potential

$$V_{eff}(r^*) := N(r)\mathcal{U}_l(\mu, M; r), r = r(r^*).$$

and hence they where characterized only by the properties of the potential. This fact is strictly true for the stationary solutions, but interestingly it was found that the quasi-stationary solutions, for which the Schrödinger-like equation no longer holds, can also be characterized by the properties of that same effective potential. Then, the cases of interest where those in which the effective potential contained a local minimum given by the condition

$$M\mu^2 r^3 - l(l+1)r^2 + 3M(l^2 + l - 1)r + 8M^2 = 0.$$

The existence of this minimum then depended solely on the combination of the parameters M_μ and l. Although none of the possible forms of the effective potential proposed allowed for bound states, the existence of the well was enough to allow for resonant states, which where the ones that where useful in constructing initial data that could give rise to long-lasting quasi-stationary configurations of finite energy.

As mentioned by the authors, it could be objected that in order to obtain the mentioned quasi-stationary solutions very particular initial data should be constructed. However, it seems that the crucial factor is the existence of the potential well. Even when starting with modified stationary solutions that are not resonant, after an initial abrupt energy loss, the late time behavior observed resulted very similar to that of the resonant quasi-stationary solutions. These solutions seemed to evolve as a combination of the resonant modes. As mentioned before, the value of μ for scalar field dark matter models is expected to be given approximately by $\hbar\mu = 10^{-24}$eV in physical units, which gives rise to effective potentials with a local minimum for all values of l.

When evaluating the characteristic times of the solutions it was found that the longest lasting configurations could last for thousands of years. Although cosmologically this is very short time, for technical reasons, the authors where only able to study cases with relatively large values of M_μ. Noting how fast the characteristic times seemed to increase with decreasing μ, it resulted unreasonable to expect that configurations with $M_\mu \sim 10^{-6}$ could last for cosmological time-scales.

Some aspects of this study are still open for some improvement. For example, a self-gravitating scalar field has not yet been considered. Second, much smaller values of the parameter μ, and much larger scalar field distributions, could be needed to do a more realistic representation of dark matter halos. The main difficulty in dealing with such configurations result in handling the very different scales numerically. Third, besides studying possible quasi-stationary or long-lasting configurations with an already existing black hole, it would also be interesting to consider more dynamical scenarios such as the formation and/or growth of the black hole and the possibility of survival of the scalar field afterwards. The results presented by Barranco et al. in [13] seem to indicate that it is indeed possible for scalar field halos around super-massive black holes to survive for cosmological time-scales.

9.5 Conclusions

In this work we have revisited an alternative DM paradigm of the Universe known as scalar field dark matter or Bose–Einstein condensate dark matter model. In this model a fundamental scalar field plays the role of dark matter.

We have reviewed a large number of recent quantitative and qualitative results aimed at explaining a variety of trends seen in the SFDM/BEC model. These trends include a brief description of Bose-Einstein condensates as dark matter, analysis and growth of its cosmological perturbations, its effect on the dynamics of galaxies, density profiles and the mass power spectrum, among others.

The key parameters of an ultralight BEC dark matter model are naturally shown to be the mass of the boson, which must be extremely small and, for the self-interacting scenario, the coupling strength of two-body repulsive interaction among the condensed particles. Dark matter is then suggested to arise from a single scalar field coupled to gravity undergoing a spontaneous symmetry break and hence rolls to a new minimum which gives a new vacuum expectation value. The breaking of symmetry can be done via a Ginzburg–Landau potential with quadratic and quartic terms. These give the mass and the interaction terms to the scalar field.

In the cosmological regime, it has been shown, that the SFDM/BEC model with an ultralight mass of 10^{-22} eV mimics the behavior of the cosmological expansion rate predicted with the ΛCDM model.

Another interesting cosmological behavior of the SF indicates that their scalar fluctuations can be appropriate for the purpose of structure formation,mainly because overdense regions of SFDM/BEC can produce the formation of galactic structure. Thus, the standard and the SFDM/BEC models can be contrasted in their predictions concerning the formation of the first galaxies. If in the future we observe more and more well formed and massive galaxies at high redshifts, this could be also a new indication in favor of the SFDM/BEC paradigm.

Much of the interest in this model has focussed on its ability to predict and agree with observations of the existence of dark matter and its capability to compete with the standard model of cosmology ΛCDM. Most of the themes we have described attempt to provide a wider view of what the model is and what is its current status. There are many other works in this field of research that are not less important, however, we trust that the interested reader will be able to deepen its knoledge with the references cited in this article.

Clearly there are a large number of competing models to describe the dark matter of the Universe, and slight differences such as vortex formation or small shifts in cosmological perturbations may not be enough to decide which one is the best, however, detail observations give us a means to discard several of them and identify the ones that can stay.

Finally, the SFDM/BEC model can have important implications in the nature of dark matter in the Universe. Additional work is needed if we are to fully understand this model, it would be desirable to have a unify framework that involves rotation curves, vortex formation, if present, and black hole effects. With all these intriguing results, we consider that the SFDM/BEC should be taken as a serious alternative to the dark matter problem in the Universe. The observational evidence seems to be in favor of some kind of cold dark matter, if we continue to observe even more galaxies at higher redshifts and if higher resolution observations of nearby galaxies exhibit a core density profile, this model can be a good alternative to ΛCDM. We expect that future observations in galaxy surveys can get us closer to the nature of dark matter.

Acknowledgments The numerical computations were carried out in the "Laboratorio de Super-Computo Astrofísico (LaSumA) del CINVESTAV", in UNAM's cluster Kan-Balam and in the cluster Xiuhcoatl from CINVESTAV. This work was partially supported by CONACyT México under grants CB-2009-01, no. 132400, CB-2011, no. 166212 and I0101/131/07 C-234/07 of the Instituto Avanzado de Cosmología (IAC) collaboration (http://www.iac.edu.mx). Abril Suárez and Victor H. Robles are supported by a CONACyT scholarship.

References

1. A. Albrecht, P.J. Steinhardt, 1982, Phys. Rev. Lett., **48**, 1220
2. M. Alcubierre, F.S. Guzmán, T. Matos et al., 2002, Class. Quant. Grav., **19**, 5017
3. A. Arbey, J. Lesgourgues, P. Salati, (2001), Phys. Rev. D, **64**, 123528
4. A. Arbey, J. Lesgourgues, P. Salati, (2002), Phys. Rev. D, **65**, 083514
5. A. Arbey, J. Lesgourgues, P. Salati, (2003), Phys. Rev. D, **68**, 023511
6. A. Arbey, (2006), EAS Publ. Ser., **20**, 257
7. A. Arbey, (2006), Phys. Rev. D, **74**, 043516
8. A. Arbey, (2009), EAS Publ. Ser., **36**, 161
9. H.W. Babcock: Lick Obs. Bulletin **19**, 41 (1939)
10. N.A. Bahcall, L.M. Lubin, V. Dorman, (1995), ApJ, **447**, L81
11. N.A.Bahcall, J.P. Ostriker, S. Perlmutter, et al., (2000), Science, **284**, 1481–1488
12. J. Bardeen, P.J. Steinhardt, M.S. Turner, (1983), Phys. Rev. D, **28**, 679
13. J. Barranco, A. Bernal, J.C. Degollado, et al., (2011), Phys. Rev. D, **84**, 083008
14. A. Bernal, F.S. Guzmán, (2006), Phys. Rev. D, **74**, 063504
15. A. Bernal, F.S. Guzmán, (2006), Phys. Rev. D, **74**, 103002
16. A. Bernal, T. Matos, D. Núnez, (2008), Rev. Mex. A. A., **44**, 149
17. J. Binney, S. Tremaine, (2008), Galactic dynamics, Princeton University Press
18. C.G. Böhemer, T. Harko, (2007), JCAP, **0706**, 025
19. A. Boriello, P. Salucci, (2001), Mon. Not. R. Astron. Soc., **323**, 285
20. M.Bradac, S.W.Allen, T. Treu, et al., 2008, ApJ, **687**, 959
21. F. Briscese, Phys. Lett. B. **696**, 315 (2011)
22. S. Burles, D. Tytler, (1998), ApJ, **499** 699
23. S. Chandrasekhar, (1939), Stellar Structure, University of Chicago Press
24. P.H. Chavanis, (2011), Phys. Rev. D, **84**, 043531
25. P.H. Chavanis, (2011), Phys. Rev. D, **84**, 043532

26. P.H. Chavanis, (2012), A & A, **537**, A127
27. D. Clowe, M. Bradac, A.H. Gonzalez, et al., (2006), ApJ, **648**, 2
28. M. Colpi, S.L. Shapiro, I. Wasserman, (1986), Phys. Rev. Lett., **57**, 2485
29. E.S. Copeland, A.R. Liddle, D. Wands, 1998, Phys. Rev. D, **57**, 4686
30. A. Cruz-Osorio, F.S. Guzmán, F.B. Lora-Clavijo, J. Cosmol. Astropart. Phys. **06**, 029 (2011)
31. M. Davis, J. Tonry, J. Huchra, et al., (1980), ApJ, **238**, L113
32. W.J.G. de Blok., arXiv:(0910).3538v1 (2010)
33. H. Dehen, B. Rose, (1993), Astrophys. Sp Sci., **207**, 133–144
34. J. Einasto, A. Kaasik, E. Saar, (1974), Nature, **250**, 309
35. S.M. Faber, J.S. Gallagher, (1979), ARA& A, **17**, 135
36. G. Fabbiano, (1989), ARAA, **27**, 87
37. W. Forman, C. Jones, W. Tucker, (1985), ApJ, **293**, 102
38. M. Gleiser, (1988), Phys. Rev. D, **38**, 2376
39. J.A. Gonzalez, F.S. Guzmán, (2011), Phys. Rev. D, **83**, 103513
40. A.X. González-Morales, A. Diez-Tejedor, L.A. Urena-López, et al., Phys. Rev. D., **87**, 021301.
41. J. Goodman, (2000), New Astronomy, **5**, 103
42. A.H. Guth, (1981), Phys. Rev. D, **23**, 347
43. A.H. Guth, S.-Y. Pi, (1982), Phys. Rev. Lett., **49**, 1110
44. A.H. Guth, P.J. Steinhardt, (1989), The Inflationary Universe, Cambridge U. Press
45. F.S. Guzmán, T. Matos, (2000), Class. Quantum Grav., **17**, L9–L16
46. F.S. Guzmán, T. Matos, H. Villegas-Brena: Astron. Nachr. **320**, 97 (1999)
47. F.S. Guzmán, L.A. Urena-López, (2003), Phys. Rev. D, **68**, 024023
48. F.S. Guzmán, L.A. Urena-López, (2004), Phys. Rev. D, **69**, 124033
49. F.S. Guzmán, L.A.Urena-López, (2006), ApJ, **645**, 814
50. S. Hanany, P. Ade, A. Balbi, et al. (2000), ApJ, **545**, L5
51. T. Harko, (2011), Phys. Rev. D, **83**, 123515
52. T. Harko, (2011), JCAP, **05**, 022
53. T. Harko, (2011), MNRAS, **413**, 3095
54. S.W. Hawking, (1982), Phys. Lett., **B115**, 295
55. J.C. Hidalgo, L. Urena-López, A. Liddle, (2012), Phys. Rev. D, **85**, 044055
56. W. Hu, R. Barkana, A. Gruzinov, (2000), Phys. Rev. Lett., **85**, 1158
57. E. Hubble, (1929), Proc. Nat. Acad. Sci., **15**, 168
58. N. Jarosik, C.L. Bennett, J. Dunkley, et al., (2010), arXiv:1001.4744
59. S.U. Ji, S.J. Sin, (1994), Phys. Rev. D, **50**, 3650
60. B. Kain, H.Y. Ling, (2010), Phys. Rev. D, **82**, 064042
61. B. Kain, H.Y. Ling, (2012), Phys. Rev. D, **85**, 023527
62. M. Yu Khlopov, B. A. Malomed, Ya. B Zeldovich, (1985), MNRAS, **215**, 575
63. A. Klypin, A.V. Kravtsov, O. Valenzuela, et al., (1999), ApJ, **522**, 82
64. T.R. Lauer, (1985), ApJ, **292**, 104
65. J. Lee, I. Koh, (1996), Phys. Rev. D, **53**, 2236
66. J.W. Lee, S. Lim, D. Choi, (2008), arXiv:astro-ph/0805.3827
67. J.W. Lee, (2009), J. Korean Phys. Soc., **54**, 2622
68. J. Lee, E. Komatzu, (2010), ApJ, **718**, 60
69. J.W. Lee, S. Lim, (2010), JCAP, **1001**, 007
70. J. Lesgourgues, A. Arbey, P. Salati, (2002), New Astron. Rev., **46**, 791
71. A.D. Lewis, D.A. Buote, J.T. Stocke, (2003), Astrophys. J., **586**, 135
72. A. Liddle, L.A. Urena-López, (2006), Phys. Rev. Lett., **97**, 161301
73. A. Liddle, C. Pahud, L.A. Urena-López, (2008), Phys. Rev. D, **77**, 121301
74. A.D. Linde, (1982), Phys. Lett., **108B**, 389
75. V. Lora, J. Magana, A. Bernal, et al., (2011), arXiv:1110.2684
76. L.M. Lubin, N.A. Bahcall, (1993), ApJ, **415**, L17
77. A.P. Lundgren, M. Bondarescu, R. Bondarescu, et al., (2010), ApJ, **715**, L35
78. J. Magana, T. Matos, A. Suárez, et al., (2012), JCAP, **10**, 003
79. D. Marsh, P. Ferreira, (2010), Phys. Rev. D, **82**, 103528
80. T. Matos, L.A. Urena-López, (2000), Class. Quant. Grav., **17**, L75
81. T. Matos, F.S. Guzmán, D. Núnez, (2000), Phys. Rev. D, **62**, 061301
82. T. Matos, L.A. Urena-López, (2001), Phys. Rev. D, **63**, 063506

83. T. Matos, A. Vázquez-González, J. Magana, (2009), MNRAS, **389**, 13957
84. T.Matos, J. Magana, A. Suárez, The Open Astron. Journal, **3**, 94.
85. T. Matos, F.S. Guzmán, arXiv: gr-qc/0108027.
86. T. Matos, L.A. Urena-López, arXiv: astro-ph/0010226.
87. B. Moore, F. Governato, J. Quinn, et al., (1998), ApJ, **499**, L15
88. B. Moore, S. Ghigna, F. Governato, et al., (1999), ApJ, **524**, L19
89. R.F. Mushotzky, H. Loewenstein, A. Awaki, et al., (1994), ApJ, **436**, L79
90. D. Núnez, J.C. Degollado, C. Palenzuela, (2010), Phys. Rev. D, **81**, 064011
91. D. Núnez, A.X. González-Morales, J.L. Cervantes-Cota, et al., (2010), Phys. Rev. D, **82**, 2
92. D. Núnez, J.C. Degollado, C. Moreno, (2011), Phys. Rev. D, **84**, 024043
93. J.P. Ostriker, J.P.E. Peebles, A. Yahil, (1974), ApJ, **193**, L1
94. T. Padmanabhan, (2003), Phys. Rept., **380**, 235
95. P.J.E. Peebles, (1982), ApJ, **263**, L1
96. P.J.E. Peebles, B. Ratra, (2003), Rev. Mod. Phys., **75**, 559
97. P.J.E. Peebles, A. Nusser, (2010), Nature, **465**, 565
98. S. Penny, C.J. Conselice, S. De Rijcke, et al., (2009), MNRAS, **393**, 1054
99. M. Perisc, P. Salucci, F. Stel, (1996), Mon. Not. R. Astron. Soc., **281**, 27
100. S. Perlmutter, G. Aldering, G. Goldhaber, et al., (1999), ApJ, **517**, 565
101. B. Ratra, P.J.E. Peebles, (1988), Phys. Rev. D, **37**, 3406
102. A. Refregier, (2003), Ann. Rev. Astron. Astrophys., **41**, 645
103. A.G. Riess, A.V. Filippenko, P. Challis, et al., (1998), Astron. J., **116**, 109
104. A.G. Riess, L-G. Strolger, J. Tonry, et al., (2004), ApJ, **607**, 665
105. A.G. Riess, L-G. Strolger, S. Casertano, et al., (2005), ApJ, **627**, 579
106. T. Rindler-Daller, P.R. Shapiro, (2011), arXiv:1106.1256v2
107. V.H. Robles, T. Matos, (2012), MNRAS, **422**, 282
108. V.H. Robles, T. Matos, (2013), ApJ, **763**, 19
109. V.H. Robles, T. Matos, (2013), arXiv:1302.5944
110. I. Rodríguez-Montoya, J. Magana, T. Matos, et al., (2010), ApJ, **721**, 1509
111. I. Rodríguez-Montoya, A. Pérez-Lorenzana, E. Cruz-Burelo De La, et al., (2011), arXiv:1110.2751
112. V.C. Rubin, W.K. Ford, N. Thonnard, (1980), Astrophys. J., **238**, 471
113. V. Sahni, L. Wang, (2000), Phys. Rev. D, **62**, 103527
114. D.N. Schramm, (1998), Nuclear and Particle Astrophysics
115. D.N. Schramm, M.S. Turner, (1998), Rev. Mod. Phys., **70**, 303
116. Z. Slepian, J. Goodman, (2012), MNRAS, **427**, 839
117. D.N. Spergel, R. Bean, O. Doré, et al., 2007, ApJ Suppl., **170**, 377
118. A.A. Starobinskii, : Phys. Lett. B. **117**, 175 (1982)
119. A. Suárez, T. Matos, (2011), MNRAS, **416**, 87
120. S.A. Thomas, F. Abdalla, O. Lahav, (2011), Phys. Rev. Lett., **106**, 241301
121. V. Trimble, (1987), ARA& A, **25**, 423
122. M.S. Turner, (1983), Phys. Rev. D, **28**, 1243
123. L.A. Urena-López, (2009), JCAP, **01**, 014
124. L.A. Urena, (2010), AIP Conf. Proc. **1318**, 82
125. H. Velter, E. Wamba, (2011), arXiv:1111.2032
126. T.P. Walket, G. Steigman, K. Schramm, et al., 1991, ApJ, **376**, 51
127. C. Wetterich: Nucl. Phys. B. **302**, 668 (1988)
128. S.D.M. White, C.S. Frenk, M. Davis, et al., (1987), ApJ, **313**, 505
129. T.-P. Woo, T. Chiueh, (2009), Astrophys. J., **697**, 850
130. N.T. Zinner, (2011), Phys. Res. Int., (2011), 734543

Chapter 10
Why We Need Dark Energy

Diego Pavón and Ninfa Radicella

Abstract If the Universe behaves as an ordinary macroscopic system, it must approach a state of maximum entropy in the long run. Realizing that the Einstein-de Sitter model cannot but dark energy dominated universes can, provided its equation of state falls in the range $-1 \leq w < -2/3$, we conclude that the present era of cosmic accelerated expansion could have been predicted on solid thermodynamic grounds. We apply these ideas to dark energy models also with variable w and to some popular modified gravity models. Further, we argue that the evolution of the Hubble function seems to hint that indeed the Universe evolves as an ordinary thermodynamic system.

10.1 Introduction

The standard cold dark matter (SCDM) model [1] was in good health until about two decades ago when it became apparent that the fractional density of matter falls well below the Einstein-de Sitter value, $\Omega_m = 1$. The death blow came at the close of the century with the discovery of the current cosmic acceleration [2], something the said model cannot accommodate by any means. However, to account for the acceleration in homogeneous and isotropic models one must either introduce some exotic energy component with a huge negative pressure (dubbed dark energy) or, more drastically, devise some theory of gravity more general than Einstein relativity [3]. Thus, both solutions appear somewhat forced and not very aesthetical. Here we first argue that dark energy (or something equivalent) is demanded on thermodynamic grounds, i.e., we provide what we believe is a sound motivation, founded on the second law, for the existence of dark energy. Then, based on the evolution of the Hubble function, we suggest that the Universe behaves as an ordinary thermodynamic system.

The natural tendency of systems to evolve toward thermodynamic equilibrium is characterized by two properties of its entropy function, $S(x)$, namely, it never decreases, $dS \geq 0$, and is convex, $d^2S < 0$ [4]. In the context of an ever expanding Friedmann-Roberson-Walker (FRW) cosmology this translates in that the entropy

D. Pavón (✉)
Departamento de Física, Universidad Autónoma de Barcelona, Barcelona, Spain,
e-mail: diego.pavon@uab.es,

N. Radicella
Dipartimento di Física, Università di Salermo, Salermo, Italy
e-mail: ninfa.radicella@uab.cat

C. Moreno González et al. (eds.), *Accelerated Cosmic Expansion*,
Astrophysics and Space Science Proceedings 38,
DOI 10.1007/978-3-319-02063-1_10,

of the apparent horizon plus that of matter and fields enclosed by it must fulfill $S'(a) \geq 0$ at any scale factor, a -the generalized second law (GSL)-, as well as $S''(a) \leq 0$ as $a \to \infty$, where the prime means d/da. The apparent horizon in FRW universes always exists (which is not generally true for the particle horizon and the future event horizon) and is known to possess not only an entropy proportional to its area [5, 6] but also a temperature [7]. Further, as demonstrated by Wang et al. [8] it appears to be the appropriate thermodynamic boundary. This set of reasons fully justifies our choice of the apparent horizon in preference to all the others. See, for instance, that the future event horizon exists just for universes that accelerate for ever, and never for decelerating universes.

Nevertheless, it is fair to recall the (at least theoretical) existence of systems lacking any global maximum entropy state, such as Antonov's sphere [9, 10]. The latter system consists in a sphere enclosing a number of particles that share some total energy. If the sphere radius happens to increase beyond some critical value, the system becomes unstable and the entropy function ceases to have a global maximum. We shall apply our argument under the assumption that the Universe tends to a state of maximum entropy irrespective of whether it may eventually reach it or not. In view of the close connection between thermodynamics and gravity -se e.g. [11–13],- it would be queer and frustrating that such a universal principle as the second law of thermodynamics would not apply to the Universe as a whole.

Section II illustrates why dark energy (or some or other modified gravity model) is required on thermodynamic basis and study some dark energy models to see whether they fulfill the thermodynamic criteria. Section III applies the said criteria to some representative modified gravity models. Finally, section IV summarizes our findings. As is customary, a naught subscript stands for the present value of the corresponding quantity.

For a wider exposition and details the reader is referred to [14] and [15].

10.2 Why Dark Energy was to be Expected

The entropy of a FRW universe is contributed by two terms: the entropy of the apparent horizon which is proportional to its area, $\mathcal{A} = 4\pi\, \tilde{r}_A^2$, and the entropy of the fluids enclosed by the horizon. Here $\tilde{r}_A = (\sqrt{H^2 + (k/a^2)})^{-1}$ denotes the radius of the horizon and H the Hubble factor of the FRW metric [5]. As is well-known, $S_A \equiv \frac{k_B}{4} \frac{\mathcal{A}}{\ell_{Pl}^2}$ where ℓ_{Pl} and k_B stand for Planck's length and Boltzmann's constant, respectively.

In virtue of the first Friedmann equation $3H^2 + 3k\, a^{-2} = 8\pi G\, \rho$, where ρ is the total energy density, we can write

$$\mathcal{A} = 4\pi\, \tilde{r}_A^2 = \frac{3}{2G}\, \frac{1}{\rho}, \qquad \text{and} \qquad \mathcal{A}' = \frac{9}{2G}\, \frac{1+w}{a\, \rho}, \tag{10.1}$$

where $w = p/\rho$ denotes the overall equation of state parameter, i.e., not just of dark energy. From the second equation in (10.1) the area will augment in expanding universes if $1 + w > 0$ and decrease otherwise.

A further derivation with w = constant yields

$$\mathcal{A}'' = \frac{9}{2G\,a^2\,\rho}\,(1 + w)(2 + 3\,w)\,. \tag{10.2}$$

Accordingly, $\mathcal{A}'' \leq 0$ for $-1 \leq w \leq -2/3$, and $\mathcal{A}'' > 0$ otherwise. Thus, the above criterion disfavor the dominance of fluids at late times such that the overall equation of state is either of phantom type or larger than $-2/3$.

We note in passing that when $w \neq$ constant last equation generalizes to $\mathcal{A}'' = 9\,(2G\,a\,\rho)^{-1}\,[w' + (1 + w)(2 + 3\,w)\,a^{-1}]$.

Let us consider the entropy associated to the fluid enclosed by the apparent horizon. If the fluid is just dust (pressureless matter, subscript m), we will have $S_m = k_B\,(4\pi/3)\,\tilde{r}_A^3\,n_0\,a^{-3} \propto a^{3/2}$. Hence, $S''_m \propto \frac{3}{4\sqrt{a}} > 0$. Notice that since $H > 0$, the entropy augments in the volume enclosed by the horizon as the latter encompasses more and more particles. Thus, $S''_m + S''_A > 0$.

Now we can be understand why sooner or later the Universe was to accelerate, i.e., why it must be endowed with dark energy (subscript x, equation of state $-1 \leq w_x \leq -2/3$), or something dynamically equivalent at the background level (as a suitably modified gravity). We note parenthetically that, at late times, dark energy is to dominate over all other energy components whereby $w \simeq w_x$ as $a \to \infty$. Were the Universe dominated by radiation and/or matter for ever, S''_A could never become negative. And if the horizon entropy dominated the total entropy, then the Universe would never tend to a state of maximum entropy (compatible with the constraints of the system). For the influence of the spatial curvature see [14].

The question now arises, whether the second derivative of the entropy of the fluid (subscript f) enclosed within the apparent horizon will be positive enough so that the sum $S''_f + S''_A$ be positive.

For simplicity we will consider $k = 0$ and a single fluid component with w_f = constant. The entropy of the fluid filling the volume enclosed by the apparent horizon can be estimated by virtue of Gibbs equation, $T_f\,dS_f = d\left(\rho_f(4\pi/3)\,H^{-3}\right) + w_f\rho_f\,d\left((4\pi/3)\,H^{-3}\right)$.

With the help of Friedmann equation (with $\rho = \rho_f$) and the equation for the evolution of the fluid temperature, $d\ln T_f/d\ln a = -3w_f$, it follows that dark energy with constant equation of state in the interval $(-1, -2/3)$ satisfies the GSL as well as the criterion that $S''_f + S''_A < 0$ when $a \to \infty$. This is consistent with the tightest observational constraints -see Table IV in [16]. In our view, cosmological models that meet both criteria should be preferred to those failing any of the two.

10.2.1 Varying *w* Models

Thus far we have focused on cosmological models in which the equation of state parameter w is a constant. Models in which the said quantity depends on time must be studied on a case by case basis. Next we consider three of such models.

10.2.1.1 Barboza and Alcaniz's Model

In this subsection we consider a phenomenological model [17] in which the dark energy equation of state is parameterized in terms of redshift (given by $1+z=a^{-1}$) as $w_x(z) = w_0 + w_1\, z(1+z)(1+z^2)^{-1}$. This expression has the advantage of not diverging at any redshift.

The pair of free, constants parameters w_0 and w_1 ought to be restricted by physical requirements and observation. A first constraint $w_0 + w_1 < 0$ follows from demanding that dark energy be subdominant at early times (when $z \gg 1$), otherwise cosmic structure would never had formed. The thermodynamic conditions of the above section constrains the said pair as

$$\begin{cases} w_1 < 2/3 \\ -1 < w_0 < -2/3 \end{cases} \cup \begin{cases} 2/3 \leq w_1 < 1 \\ -1 < w_0 < -w_1, \end{cases} \tag{10.3}$$

Interestingly enough, this result is in full agreement with the observational constraints obtained by using data from supernovae type Ia, baryon acoustic oscillations, and cosmic microwave background, namely: $-1.35 \leq w_0 \leq -0.86$, $-0.33 \leq w_1 \leq 0.91$ [17].

10.2.1.2 Chaplygin Gas Model

The original Chaplygin model unifies matter and dark energy in the sense that they are no longer two separate components but a unique entity that mimics cold matter at early times and a cosmological constant at late times [18]. Its equation of state $p = -A/\rho$, with A a positive constant and ρ the energy density of the gas (i.e., the total energy density), is obtainable from the Nambu-Goto action for a d-brane moving in the $d+1$ dimensional bulk [19]. In a FRW universe, the dependence of the energy density on the scale factor reads $\rho = \sqrt{A + (B/a^6)}$, where B is nonnegative integration constant.

It is readily seen that $S'_{\mathcal{A}}$ is positive at all times and $S''_{\mathcal{A}}$ negative at late times. On its part, the Chaplygin entropy obeys $S'_{Ch} < 0$ and $S''_{Ch} > 0$ for large scale factor. However, as it can be checked, $S'_{Ch}/S'_{\mathcal{A}} \to 0$ and $S''_{Ch}/S''_{\mathcal{A}} \to 0$ as $a \to \infty$. Accordingly, the total entropy, $S = S_{Ch} + S_{\mathcal{A}}$, fulfills $S' > 0$ and $S'' < 0$ in the same limit. In other words, the GSL is satisfied as $a \to \infty$ and the Chaplygin universe tends to thermodynamic equilibrium in the long run.

10.2.1.3 Holographic, Interacting Models

Here we first consider a spatially-flat, holographic, interacting model dominated by pressureless matter and dark energy. The latter component is assumed holographic in the sense that its energy density varies as the area of the apparent horizon and interacts with matter at a given, non-constant, rate. As a result w_x decreases with expansion and the fractional densities of both components remain constant. This much alleviates the cosmic coincidence problem [20] and the model shows compatibility with observation [21].

Inspection of the Hubble factor, given by Eq. (2.4) of [21],

$$H(a) = H_0 \left[\frac{\Gamma}{3H_0\, r} + \left(1 - \frac{\Gamma}{3H_0\, r}\right) a^{-3/2} \right], \tag{10.4}$$

where $r = \rho_m/\rho_x =$ constant, readily reveals that the graph of the evolution of the area of the apparent horizon, $\mathcal{A} \propto 1/H^2$, is growing, has positive curvature at the beginning of the expansion and gently evolves to negative curvature values remaining thus for ever for the best fit observational values, $\Gamma/H_0 = 0.563$ and $r = 0.452$, obtained in [21].

As can be readily checked, the second derivative of the entropy of the dark energy within the horizon, S_x, is negative as well, while the entropy of matter increase with expansion and its second derivative is positive at early times and negative at late times. Thus, the GSL is fulfilled and $S''_A + S''_m + S''_x < 0$ when $a \to \infty$.

However, it would be naïve to believe that for all holographic models $\mathcal{A}'' < 0$ when $a \gg 1$. For instance, in the model of Gao et al. [22], which uses the Ricci's length as infrared cutoff, the Hubble function takes the form

$$H = H_0 \sqrt{\Omega_{k0}\, a^{-2} + \Omega_{m0}\, a^{-3} + \Omega_{r0}\, a^{-4} + \alpha(2-\alpha)^{-1}\, \Omega_{m0}\, a^{-3} + f_0\, a^{(2/\alpha)-4}}\,,$$

where the subscript r stands for radiation; $\alpha \simeq 0.46$ and $f_0 \simeq 0.65$ are dimensionless parameters. In this case, the area of the apparent horizon increases to a maximum to monotonously decrease forever afterwards. Consequently, $\mathcal{A}'' > 0$ as $a \to \infty$.

On the other hand, the entropies of the fluid components (radiation, matter, and dark energy) go down with expansion and their second derivatives are positive for large scale factor. Thus, this model violates the GSL and does not approach thermodynamic equilibrium at late times.

10.3 Modified Gravity Models

Models that depart from Einstein gravity may lead to a late acceleration era without the help of any exotic component of negative pressure. Here we examine some of them, namely: the one based in the brane-induced gravity model of Dvali et al. [23], the original Cardassian model proposed by Freese and Lewis [24], and the torsion model of Bengochea and Ferraro [25], to check whether they fulfill these criteria.

10.3.1 Dvali-Gabadazze-Porrati's Model

This model considers our 4-dimensional Universe as a brane embedded in a 5-dimensional bulk with flat Minkowski metric. As a consequence of the brane-induced term, the conventional Friedmann's equation modifies to

$$H^2 + \frac{k}{a^2} = \left(\sqrt{\frac{\rho}{3M_{Pl}^2} + \frac{1}{4r_c^2}} + \frac{1}{2r_c} \right)^2, \tag{10.5}$$

where ρ and r_c stand for the total energy density (matter plus radiation in this model) and the crossover scale below which gravity appears as four dimensional, respectively.

A detailed study shows that the GSL is satisfied and that $S''_A + S''_r + S''_m < 0$ at late times provided the present number density of dust particles is bounded by

$$n_0 < \frac{1}{k_B}\left[\frac{27 k_B \Omega_{m0} r_c H_0^2}{4\,\ell_{Pl}^2 c}\right] \sim 10^{38}\ \text{cm}^{-3}\,, \tag{10.6}$$

which is true by a huge margin.

10.3.2 Cardassian Model

In this spatially-flat FRW model the first Friedmann equation acquires an extra term that accounts for acceleration at sufficiently high redshifts, $H^2 = (8\pi G/3)\,\rho + B\rho^{\alpha}$. Here two new non-negative constants, B and α, appear while ρ stands for the energy density of cold matter -the only energy component. This model can be mapped to a dark energy model via the identifications $\alpha = 1 + w_x$ and $B = (8\pi G/3)\,(\rho_{x0}/\rho_{m0}^{\alpha})$. It thus follows that for $0 < \alpha < 2/3$ one has $S'_m + S'_A > 0$ and $S''_m + S''_A < 0$ as $a \to \infty$.

10.3.3 Torsion Model

This class of models describes gravitation in terms of the torsion scalar τ rather than curvature. The action takes the form

$$I = \frac{1}{16\pi G}\int d^4x\sqrt{-g}\,(\tau + f(\tau)) + I_{matter}\,, \tag{10.7}$$

where $f(\tau)$ is a free function to be constrained by observation and experiments. Here we focus in the proposal of Bengochea and Ferraro [25] where $f(\tau) = -\alpha(-\tau)^{-n}$ with $\alpha = (1 - \Omega_{m0})(6H_0^2)^{1-n}/(2n-1)$. A detailed analysis reveals that neither the GSL is respected nor the total entropy is convex in the long run. Therefore, this specific model is not thermodynamically sound.

10.4 Does the Universe Behaves as an Ordinary System?

Thus far we have assumed that the Universe as a whole tends to some or other thermodynamic equilibrium state in the long run. Put another way, that its entropy never decreases and that it is convex at least in the last stage of approaching equilibrium (if it were positive such state would not be reachable). Here, based on the available observational data on the evolution of the Hubble function, we argue that our assumption is more likely to be correct than otherwise [15] (Fig. 10.1).

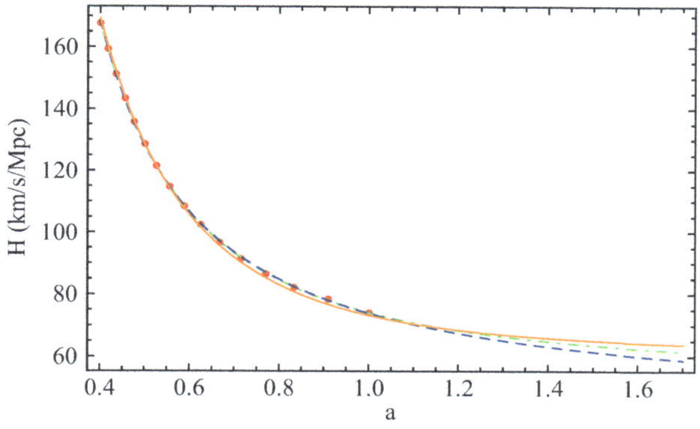

Fig. 10.1 The string of points show the Hubble history in the interval $0.4 \leq a \leq 1$. The one at $a = 1$ indicates the Hubble constant value, H_0, as measured by Riess et al. [35]. The other fourteen points correspond to simulated values of the Hubble function assuming an accuracy of 1 % in the $H(a)$ observations according to Carvalho and Alcaniz (Fig. 3a) in Ref. [34]). The dashed, dot-dashed, and solid lines are the best fit curves of the models represented by Eq. (10.10.1), Eq. (10.10.2), and the spatially flat ΛCDM model, respectively

We first note that the said observations strongly suggest that $H'(a) < 0$ and $H''(a) > 0$, which is fully consistent recent studies on the impact of hypothetical transient periods of acceleration-deceleration on the matter growth [26] and on the radiation power spectrum [27] from the decoupling era, $a \simeq 10^{-5}$, to $a = 0.5$, and with the study of Serra et al. [28]. The latter shows that equation of state parameter of dark energy has not been noticeably varied between $a = 0.5$ and 1, where the normalization $a_0 = 1$ is understood. These studies firmly point to the absence of the said hypothetical periods. A further and crucial observation is that the current value of the deceleration parameter is negative, $q_0 = -[1 + (aH'/H)]_0 < 0$ -see e.g., [2, 29, 30]. In addition, measurements of the Hubble factor at different redshifts [31, 32] plus numerical simulations [33, 34] fairly suggest that $H'(a) < 0$ and $H''(a) > 0$, at least in the interval $0.4 \leq a \leq 1$ -see Figs. 1c and 3a in [34], and Fig. 10.1. The latter shows the projected evolution of the Hubble function in terms of the scale factor in the said interval; the set of points was adapted from Fig. 3a of [35], which results from numerical simulations -assuming a precision of 1 %- of H(z) measurements from luminous red galaxies [33], plus the recently measured value of the Hubble constant, $H_0 = 74.2 \pm 3$ Km/s/Mpc -see Riess et al. [35].

Arguably, the entropy of FRW universes is dominated by the entropy of the causal horizon, at least at late times -see e.g. [36]. As causal horizon we take the apparent horizon. In the case of the spatially flat FRW metric its area reduces to $\mathcal{A} = 4\pi H^{-2}$. Thus,

$$\mathcal{A}' \propto -\frac{2\,H'}{H^3}, \qquad \text{and} \qquad \mathcal{A}'' \propto \frac{2}{H^2}\left[3\left(\frac{H'}{H}\right)^2 - \frac{H''}{H}\right]. \tag{10.8}$$

According to the above, the second law of thermodynamics imposes that $\mathcal{A}' \geq 0$, at all times, and that $\mathcal{A}'' < 0$ -at least as $a \to \infty$.

The overall behavior the aforesaid figures show is shared by the spatially flat ΛCDM model which seems to pass fairly well most, if not all, observational tests. This implies that whatever the "right" cosmological model turns out to be, it will not substantially differ, observationally, from the ΛCDM. Since there is no apparent reason for this trend to change in the future (it would if the Universe expansion were dominated by phantom dark energy, but phantom models are plagued with quantum instabilities [37, 38] and can be dismissed) we shall assume that the inequalities of above will stay in place also for $a > 1$. In consequence $\mathcal{A}'$ will result positive-definite, however $\mathcal{A}''$ may bear any sign. By imposing that $\mathcal{A}''$ should be negative, the constraint

$$3\left(H'/H\right)^2 < H''/H \tag{10.9}$$

readily follows.

From the set of conditions $H' < 0$, $H'' > 0$, and $q < 0$ -the latter holding only from some "recent time" on-, it can be demonstrated that for sufficiently large scale factor onwards the inequality (10.9) is to be satisfied and, accordingly, $\mathcal{A}'' < 0$. Effectively, bear in mind that $q = -[1 + (aH'/H)]$; then $H'/H = -(1+q)/a$. Since H' and q are negative the numerator of last expression stays bounded (it lies in the range $0 \leq 1+q \leq 1$), whence the left hand side of (10.9) vanishes in the long run.

Inspection of panels of Fig. 1c and Fig. 3d in [34], as well as Fig. 10.1, suggests that the data points can be roughly approximated by the simple expressions

$$H = H_* \exp(\lambda/a) \qquad \text{and} \qquad H = H_*(1 + \lambda a^{-n}), \tag{10.10}$$

where $H_* = H(a \to \infty) > 0$, $\lambda > 0$, and $n > 1$. Both functions describe ever expanding universes with $H' < 0$ and $H'' > 0$. By inserting the first one in (10.9) one obtains that $\mathcal{A}'' < 0$ from the instant the Universe starts accelerating onwards, namely, for $a \geq \lambda$. By fitting (10.10.2) to the set of points displayed in Fig. 10.1 (dashed line) we find that $H_* = 42.6 \pm 0.4$ km/s/Mpc and $\lambda = 0.550 \pm 0.005$, both at 95 % confidence level (CL). For the cosmic expansion described by (10.10.2) the best fit values, at 95 % CL, of the parameters are $H_* = 54.78 \pm 0.06$ km/s/Mpc, $\lambda = 0.3535 \pm 0.0013$ and $n = 1.928 \pm 0.002$ (dot-dashed line in Fig. 10.1).

Nevertheless, it is to be noted that not all Hubble functions that fulfill the observational restrictions $H' < 0$ and $H'' > 0$ comply with the inequality (10.9). This is, for instance, the case of the expansion laws $H = H_*[\exp(\lambda a^{-1}) - 1]$ and $H = H_* \exp(-\lambda a)$ (with $\lambda > 0$). Clearly, the entropy of a universe that obeyed any of these two laws would increase without bound in the long run, similarly to the entropy of Antonov's sphere in Newtonian gravity. Note, however, that the said functions do not correspond to realistic universes. In the first case the universe never accelerates; in the second one it accelerates at early times (when $a < 1/\lambda$) to decelerate forever afterwards. Thus, neither of them results compatible with observation. This suggest that Hubble functions that satisfy the inequalities $H'(a) < 0$ and $H''(a) > 0$ but violate Eq. (10.9) (thereby leading to an unbound entropy as $a \to \infty$) are unrealistic.

10.5 Conclusions

If the Universe really behaves as a ordinary thermodynamic system and can be described at large scales by the FRW metric, then at some point of its evolution it must experience a never ending period of accelerated expansion. Therefore, its present accelerated state could have been predicted on solid physical grounds well before its experimental discovery. Nevertheless, it remains the question as to whether the Universe fulfills the second law of thermodynamics. As argued in the previous section, the answer is more likely to be yes than not. A definitive answer requires a rather large increase in the amount and quality of useful data.

References

1. P.J.E. Peebles, *Principles of Physical Cosmology* (Princeton University Press, Princeton, NJ, 1993)
2. A.G. Riess et al., Astron. J. **116**, 1009 (1998); S. Perlmutter et al., Astrophys. J. **517**, 565 (1999)
3. R. Durrer, R. Maartens, Gen. Relativ. Grav. **40**, 301 (2008); L. Perivolaropoulos, AIP Conf. Proc. **848**, 698 (2006); J. Frieman, M. Turner, D. Huterer, Ann. Rev. Astron. Astrophys. **46**, 385 (2008); R.R. Caldwell, M. Kamionkowski, Ann. Rev. Nucl. Part. Sci. **59**, 397 (2009); L. Amendola, S. Tsujikawa, Dark Energy, (CUP, Cambridge, 2010)
4. H.B. Callen, *Thermodynamics* (J. Wiley, New York, 1960)
5. D. Bak, S.-J. Rey, Class. Quantum Grav. **17**, L83 (2000).
6. R.-G. Cai, Progr. Theoret. Phys. Suppl. **172**, 100 (2008)
7. R.-G. Cai, L.-M. Cao, Y.-P. Hu, Class. Quantum Grav. **26**, 155018 (2009)
8. B. Wang, Y. Gong, E. Abdalla, Phys. Rev. D. **74**, 083520 (2006)
9. D. Lynden-Bell, R. Wood, Mon. Not. R. Astron. Soc. **138**, 495 (1968)
10. D. Lynden-Bell, Physica A. **263**, 293 (1999)
11. S.W. Hawking, Commun. Math. Phys. **43**, 199 (1975)
12. T. Jacobson, Phys. Rev. Lett. **75**, 1260 (1995)
13. T. Padmanbhan, Phys . Rep. 406, 49; T. Padmanabhan,www.gravityresearchfoundation.org/pdf/awarded/2008/Padmanabhan_2008.pdf (2005)
14. Ninfa Radicella and Diego Pavòn, Gen. Relativ. Gravit. **44**, 685 (2012)
15. Ninfa Radicella and Diego Pavòn, Gen. Relativ. Gravit. **45**, 63 (2013)
16. E. Komatsu et al., Astrophys. J. Supplement. **192**, 18 (2011).
17. E.M. Barboza Jr., J.S. Alcaniz, Phys. Lett. B. **666**, 415 (2008).
18. A.Y. Kamenschchik, U. Moschella, V. Pasquier, Phys. Lett. B. **511**, 265 (2001).
19. R. Jackiw, arXiv:physics/0010042
20. W. Zimdahl, D. Pavón, Class. Quantum Grav. **24**, 5461 (2007)
21. I. Durán, D. Pavòn, W. Zimdahl, JCAP **07** 018 (2010)
22. C. Gao, F. Wu, X. Chen, and Y.G. Shen, Phys. Rev. D. **79**, 043511 (2009)
23. G. Dvali, G. Gabadadze and M. Porrati, Phys. Lett. B. **485**, 208 (2000)
24. K. Freese and M. Lewis, Phys. Lett. B. **540**, 1 (2002)
25. G.R. Bengochea, R. Ferraro, Phys. Rev. D. **79**, 124019 (2009)
26. E.V. Linder, Phys. Rev. D. **82**, 063514 (2010)
27. E.V. Linder, T.L. Smith, JCAP **04**, 001 (2011)
28. P. Serra et al., Phys. Rev. D. **80**, 121302(R) (2009)
29. R. Daly et al., Astrophys. J. **677**, 1 (2008).
30. R. Amanullah et al., Astrophys. J. **716**, 712 (2010)
31. J. Simon et al., Phys. Rev. D. **71**, 123001 (2005)

32. D. Stern et al., JCAP **02**, 008 (2010)
33. S.M. Crawford et al., Mon. Not. R. Astron. Soc. **406**, 2569 (2010)
34. J.C. Carvalho, J.S. Alcaniz, Mon. Not. R. Astron. Soc. **418**, 1873 (2011)
35. A.G. Riess et al., Astrophys. J. **699**, 539 (2009)
36. C.A. Egan, C.H. Lineweaver, Astrophys. J. **710**, 1825 (2010)
37. S.M. Carroll, M. Hoffman, M. Trodden, Phys. Rev. **D68**, 023509 (2003)
38. J.M. Cline, S. Jeon, G.D. Moore, Phys. Rev. **D 70**, 043543 (2004)

Chapter 11
Non-Spherical Voids: the Best Alternative to Dark Energy?

Roberto A Sussman

Abstract The constraints from current cosmological observations strongly support the ΛCDM model in which late time cosmic dynamics is dominated by a nonzero cosmological constant or by an exotic and elusive source like "dark energy". However, these constraints can also be met if we assume a non-perturbative treatment of cosmological inhomogeneities and that our location lies within an under–dense or "void" region of at least 300 Mpc characteristic length. Since fitting observational data severely constrains our position to be very near the void center in spherical void models, we propose in this article a toy model of a less idealized non-spherical configuration that may fit this data without the limitations associated with spherical symmetry. In particular, the class of quasi–spherical Szekeres models provides sufficient degrees of freedom to describe the evolution of non-spherical inhomogeneities, including a configuration consisting of several elongated supercluster-like overdense filaments with large underdense regions between them. We summarize a recently published example of such configuration, showing that it yields a reasonable coarse-grained description of realistic observd structures. While the density distribution is not spherically symmetric, its proper volume average yields a spherical density void profile of 250 Mpc that may be further improved to agree with observations. Also, once we consider our location to lie within a non-spherical void, the definition of a "center" location becomes more nuanced, and thus the constraints placed by the fitting of observations on our position with respect to this location become less restrictive.

11.1 Introduction

Inhomogeneous cosmological models have become a valuable tool to analyze cosmological observations without introducing an elusive dark energy source (a comprehensive review on this is found in [1]). The currently preferred inhomogeneous configurations are Gpc-scale under–densities ("voids") based on the spherically symmetric Lemaître-Tolman (LT) models [2, 3], under the assumption that we live close

R. A Sussman (✉)
Instituto de Ciencias Nucleares, UNAM, México,
Circuito Exterior C.U., 04510 México,
e-mail: sussman@nucleares.unam.mx

C. Moreno González et al. (eds.), *Accelerated Cosmic Expansion*,
Astrophysics and Space Science Proceedings 38,
DOI 10.1007/978-3-319-02063-1_11,

to a center of a cosmic density depression of radius around 1–3 Gpc [4, 5, 6]. Criticism has been voiced on these void models on the grounds that they violate the Copernican principle, since compliance with the cosmic microwave background (CMB) constraints allows for only one such Gpc structure and the observer location cannot be further away from the origin than $\sim$ 50 Mpc [7] (see also [6]). However, as suggested by more recent work [8, 9], a void of radius 250 Mpc may be sufficient to explain the supernova observations, the power spectrum of the CMB and is also consistent with Big Bang Nucleosynthesis, or Baryon Acoustic Oscillations. By considering void structures of this size the Copernican Principle is not violated, as our Universe may consist of many such structures (the upper size to violate CMB constrains is 300 Mpc [10, 11]). Evidently, restricting our position to be within 50 Mpc from the center origin of a 250 Mpc void is a less stringent limitation. Notice that these voids are not the smaller voids (30–50 Mpc) seen in the filamentary structure of our Local Universe that roughly correspond to numerical simulations, but would form a structure a larger voids containing the smaller ones yet to be detected by observations.

In a recent article [12] we examined the possibility of using non–spherical void models to describe cosmic inhomogeneities. For this purpose, we considered the class of non–spherical Szekeres solutions of Einstein's equations [13, 14, 15, 16]. By fixing the free parameters of these solutions by means of a thin-shell approximation [10, 11, 17, 18], we obtained a specific model that yields a reasonable coarse-grained description of realistic cosmic structures. Since we define initial conditions at the last scattering surfaces, this model evolves from small early universe initial fluctuations and is consistent with current structure formation scenarios. The model presented in [12] yields an averaged spherically symmetric density distribution with a radial void profile qualitatively analogous to the spherical void models (as those of [8]), hence suggesting that the latter models may be approximate configurations that should emerge after coarse-graining and averaging of under–dense regions of a realistic lumpy non–spherical Universe. Also, the lack of spherical symmetry in the Szekeres model removes the unique invariant nature of the center location of models with this symmetry. Since our being sufficiently near this center is a strong constraint that the fitting of observations place on spherical LT models, this constraint becomes much less restrictive in a non-spherical Szekeres model.

11.2 Setting up the Szekeres Model

The metric of Szekeres models takes the following form [13]

$$ds^2 = dt^2 - \frac{(\Phi' - \Phi\mathcal{E}'/\mathcal{E})^2}{\epsilon - k} dr^2 - \frac{\Phi^2}{\mathcal{E}}(dx^2 + dy^2), \tag{11.1}$$

where $\Phi = \Phi(t, r)$ and $\Phi' = \partial\Phi/\partial r$, with:

$$\mathcal{E} = \frac{S}{2}\left[\left(\frac{x-P}{S}\right)^2 + \left(\frac{y-Q}{S}\right)^2 + \epsilon\right], \tag{11.2}$$

while $k(r), S(r), P(r), Q(r)$ are arbitrary functions; ϵ is a constant: the values $\epsilon = 1, 0, -1$ are respectively known as the quasi–spherical, quasi–plane and quasi-hyperbolic Szekeres models (for a detailed discussion on these models see [14, 15, 16]). We consider only the quasispherical case, in which the surfaces marked by r and t constant can be mapped to 2–spheres by a stereographic projection.

Einstein's equations for a dust source associated with (11.1)–(11.2) reduce to

$$\dot{\Phi}^2 = -k(r) + \frac{2M(r)}{\Phi}, \tag{11.3}$$

$$8\pi G\rho = \frac{2M' - 6M\mathcal{E}'/\mathcal{E}}{\Phi^2(\Phi' - \Phi\mathcal{E}'/\mathcal{E})}, \tag{11.4}$$

where $M(r)$ is an arbitrary function and we assume that $\Phi' \neq \Phi\mathcal{E}'/\mathcal{E}$ holds whenever $M' \neq 3M\mathcal{E}'/\mathcal{E}$, in order to avoid a shell crossing singularity [16, 26]. The solution of (11.3) is given by the quadrature

$$\int_0^{\Phi} \frac{\mathrm{d}\widetilde{\Phi}}{\sqrt{-k + 2M/\widetilde{\Phi}}} = t - t_B(r). \tag{11.5}$$

where $t_B(r)$ marks the locus of the big bang (which is, in general, non–simultaneous). We remark that this model has no isometries (it does not admit Killing vectors), but by specializing the free functions we obtain axially and spherically symmetric models as particular cases.

By choosing the r coordinate such that $\bar{r} = \Phi(t_i, r)$, where $t = t_i$ marks the last scattering surface (and dropping the bar to simplify notation), we can eliminates one of the six independent functions of r appearing above. Thus, in order to achieve with a Szekeres model the most realistic possible description of cosmic structures and structure formation, we must prescribe five free functions as initial conditions to specify a unique model. In particular, we will specify the functions S, P, Q, t_B and M. The algorithm that we use in the calculations can be defined as follows:

1. The chosen asymptotic cosmic background is an open Friedman model[1], i.e. $\Omega_m = 0.3$ and $\Lambda = 0$. The background density is then given by

$$\rho_b = \Omega_m \times \rho_{cr} = 0.3 \times \frac{3H_0^2}{8\pi G}(1+z)^3, \tag{11.6}$$

   where the Hubble constant is $H_0 = 70$ km s^{-1} Mpc^{-1}.
2. We choose $t_B = 0$, hence the age of the Universe (given by (11.5)) is everywhere the same (as in the homogeneous background Friedmann model) and is equal to $t_i = 471{,}509.5$ years (see [20] for details).

[1] Asymptotic spatial flatness is no longer required if homogeneity is relaxed [6, 19].

3. The function $M(r)$ is given by

$$M(r) = 4\pi \frac{G}{c^2} \int_0^r \rho_b(1 + \delta\bar{\rho})\,\bar{r}^2\,\mathrm{d}\bar{r},$$

where $\delta\bar{\rho} = -0.005\mathrm{e}^{-(\ell/100)^2} + 0.0008\mathrm{e}^{-[(\ell-50)/35]^2} + 0.0005\mathrm{e}^{-[(\ell-115)/60]^2} + 0.0002\mathrm{e}^{-[(\ell-140)/55]^2}$, and $\ell \equiv r/1$ kpc.
4. The function $k(r)$ can be calculated from (11.5).
5. The functions Q, P, and S are prescribed in order to provide the best possible coarse–grained description of the density distribution of our observed local Cosmography by means of a thin shell approximation (see [12]).
6. Once the model is specified, its evolution is calculated from Eq. (11.3) and the density distribution at the current instant is evaluated from (11.4).

11.3 How Realistic this Model Can be?

The density distribution for our model (depicted in Fig. 11.1 in intuitive Cartesian coordinates [21, 22]) follows from our choice of the functions $\{M, t_B, Q, P, S\}$. If new data would arise showing a different density pattern, we can always adjust it appropriately by selecting different functions that would change the position, size, and the amplitude of the overdensities (see [21, 22] for a detailed discussion).

As shown in Fig. 11.1, the model under consideration contains structures such as voids and elongated supercluster–like overdensities. It has large overdensities around $\sim$ 200 Mpc (towards the left of the) that compensate the underdense regions and allow the model to be practically homogeneous at $r > 300$ Mpc. Actual observations reveal very massive matter concentrations – the Shapley Concentration roughly at the distance of 200 Mpc, or the Great Sloan Wall at the distance of 250–300 Mpc. In the opposite direction on the sky we find the Pisces–Cetus and Horologium–Reticulum, which are massive matter concentrations located at a similar distance. We refer the reader to Fig. 44 of Ref. [24], which provides a density map of the Local Universe reconstructed from the 2dF Galaxy Redshift Survey Survey using Delaunay Tessellation Field Estimator[2]. Also, the inner void seen in Fig. 11.1 is consistent with what is observed in the Local Universe – it appears that our Local Group is not located in a very dense region of the Universe, rather it is located in a less dense region surrounded by large overdensities like the Great Attractor on one side and the Perseus–Piscis supercluster on the other side. Both are located at around 50 Mpc—see Fig. 19 of [25] that provides the density reconstruction of the Local Universe using the POTENT analysis.

While still far from a perfect "realistic" description, the density pattern displayed in Fig. 11.1 exhibits the main features of our local Universe. It should be therefore treated as a "coarse-grained" approximation to study local cosmic dynamics by means of a suitable exact solution of Einstein's equations. Such approximation is, evidently,

[2] This figure is also available at http://en.wikipedia.org/wiki/File:2dfdtfe.gif

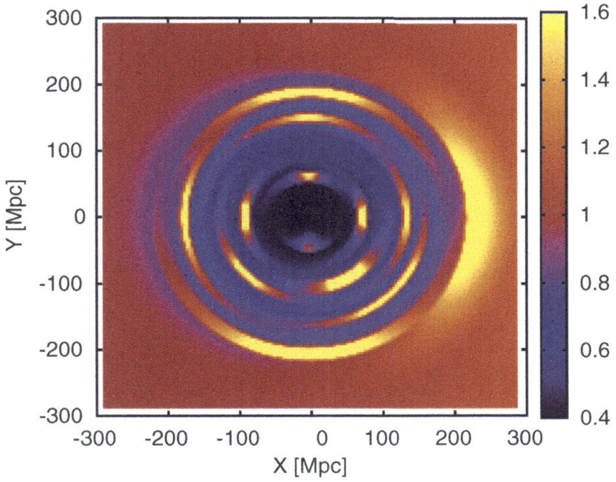

Fig. 11.1 The present-day color-coded density distribution ρ/ρ_0 (where ρ_0 is density of the homogeneous background model). Brighter colors indicate a high-density region, darker low-density region 11.3

far less idealized than the gross one that follows from spherically symmetric LT models.

11.4 Position of the "Center"

As a consequence of the lack of spherical symmetry, the model under consideration lacks an invariant and unique characterization of a center worldline. Instead, for every 2–sphere corresponding to a fixed value of r at an instant t = constant, we have (at least) two locations that can be considered appropriate generalizations of the spherically symmetric center: the worldline marked by the coordinate "origin" $r = 0$ where the shear tensor vanishes, which defines a locally isotropic observer (cf. eq (16.29) of Ref. [26]), and the "geometric" center of the 2–sphere whose surface area is $4\pi\Phi^2$.

As shown in Fig. 11.2, the fact that the 2–spheres of constant r in a quasi-spherical Szekeres model are non-concentric implies that the geometric center of these spheres and $r = 0$ do not coincide. As a consequence, the distance from this origin to the surface of the sphere depends on the direction marked by the angles (θ, ϕ) of the stereographic projection (see Eq. (3) of Ref. [12]):

$$\delta(r, \theta, \phi) = \int_0^r \mathrm{d}\tilde{r} \frac{\Phi' - \Phi\mathcal{E}'/\mathcal{E}}{\sqrt{1-k}}. \qquad (11.7)$$

Hence, the displacement Δ between the origin and the geometric center of a sphere of radius r is

$$\Delta = \frac{\delta_{\max} - \delta_{\min}}{2},$$

Fig. 11.2 Schematic representation of locations that can be considered “centers” in a quasi–spherical Szekeres model: the local isotropic observer at the origin $r = 0$ (denoted by a black dot) where shear vanishes and the geometric center of the larger sphere depicted by a cross. The distance between these locations is denoted by Δ.

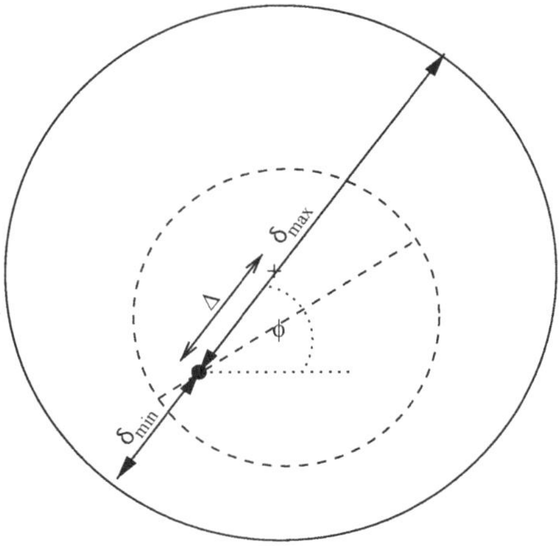

where $\delta_{\rm max} = \max(\delta)$, $\delta_{\rm min} = \min(\delta)$. As can be seen from Eq. (3) of [12], the maximal and minimal value of $\mathcal{E}'/\mathcal{E}$ for our model (where $S' = 0$) corresponds to $\theta = \pi/2$. The distance, δ, as a function of ϕ for voids of various radii is depicted by Fig. 3 of [12], showing that a sphere whose present-day area radius is $\Phi = 100$ Mpc the model under consideration yields a displacement of $\Delta = 36$ Mpc towards $\phi \approx 80°$ direction. While for $\Phi = 250$ Mpc we have $\Delta = 62$ Mpc towards $\varphi \approx 120°$.

Fitting observations in spherically symmetric models restricts our cosmic location to be within a given maximal separation from a location that is both, the geometric center of the void and the locally isotropic observer ($\Delta = 0$). It is reasonable to expect that similar distance restrictions with respect to the local isotropic observer should emerge in fitting observations with a Szekeres model, but in the latter models this observer is not the only center and may be far away from the geometric center of the void, and thus our location would be less special and improbable than in spherically symmetric models where both locations coincide.

11.5 Averaging

As shown in Ref. [23], the proper 3–dimensional volume in space slices orthogonal to the 4–velocity (t = constant) in a Szekeres model is

$$V_{\mathcal{D}} = \int_0^{r_{\mathcal{D}}} \mathrm{d}r \int_{-\infty}^{\infty} \mathrm{d}x \int_{-\infty}^{\infty} \mathrm{d}y \sqrt{-g} = 4\pi \int_0^{r_{\mathcal{D}}} \mathrm{d}r \frac{\Phi^2 \Phi'}{\sqrt{1-k}} \equiv 4\pi R_{\mathcal{D}}, \tag{11.8}$$

and thus, the proper volume averaged density is spherically symmetric (*i.e.* independent of x and y), even if the density itself is far from a spherical distribution:

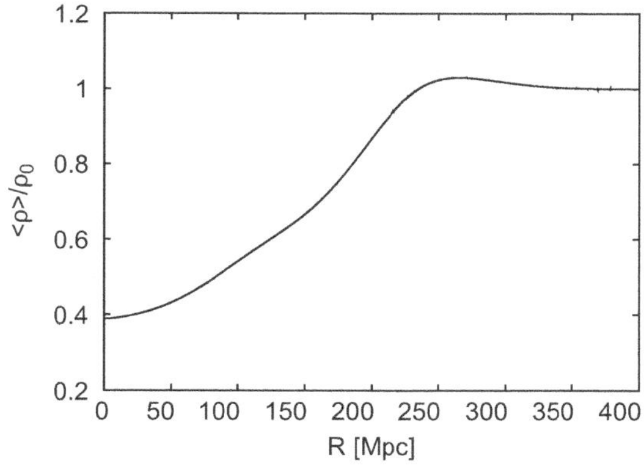

Fig. 11.3 Radial profile of the spherically symmetric averaged distribution (normalized by the background density ρ_0)

$$\langle\rho\rangle(r_{\mathcal{D}}) = \frac{1}{V_{\mathcal{D}}}\int_0^{r_{\mathcal{D}}} \mathrm{d}r \int_{-\infty}^{\infty} \mathrm{d}x \int_{-\infty}^{\infty} \rho\, \mathrm{d}y \sqrt{-g}$$
$$= \frac{1}{8\pi G R_{\mathcal{D}}}\int_0^{r_{\mathcal{D}}} \mathrm{d}r \frac{2M'}{\sqrt{1-k}}. \qquad (11.9)$$

The radial profile of this spherical volume–averaged density distribution evaluated as a function of $r_{\mathcal{D}}$, is displayed by Fig. 11.3. The spherical symmetry of the averaged density distribution implies that the the averaging process has smoothed out the "angular" (*i.e.* x, y) dependence of a highly non–spherical coarse grained density distribution. Since the resulting averaged distribution $\langle\rho\rangle(r_{\mathcal{D}})$ is equivalent to a spherical cosmic void whose radius is approximately 250 Mpc (as in Ref. [8]), the latter type of void models can be thought of as rough averages of more realistic non–spherical configurations. As a consequence, the use of a Szekeres model seems to suggest that results obtained by means of spherical LT models may be robust: while local non–spherical information could still provide important refinements, and is needed for computations involving null geodesics (specially when fitting CMB constraints), it is likely that basic bottom line information is already contained in the spherical voids constructed with LT models.

11.6 Conclusions

The model we have presented is one among the first attempts in using the Szekeres solution as a theoretical and empiric tool to study and interpret cosmological observations [28, 29, 9, 30]. This opens new possibilities for inhomogeneous cosmologies, as this is the most general available cosmological exact inhomogeneous and anisotropic solution of Einstein's equations. The model provides a more nuanced and much less restrictive description of the need to constrain our location with respect to a center location. It is also a concrete example that illustrates the possibility that a mildly increasing void profile (required by observations) can emerge if local structures are coarse-grained and then averaged. Of course, notwithstanding these appealing features, the model and its assumptions must be subjected to hard testing by data from the galaxy redshift surveys, and evidently the more comprehensive this data can be the better it can be used for this purpose. Unfortunately current surveys like 2 dF of SDSS do not cover the whole sky and only focus on small angular regions of it. However in the near future this limitation may be overcome – for example, Sky Mapper[3] aims to cover the whole southern sky which will provide sufficient data to test possibilities suggested and elaborated in this work. A more comprehensive and detailed article on the model proposed here is currently under elaboration and will be submitted soon for publication.

References

1. K. Bolejko et al., *Structures in the Universe by Exact Methods—Formation, Evolution, Interactions*, Cambridge University Press, Cambridge, 2009
2. G. Lemaître, Ann. Soc. Sci. Bruxelles A53 (1933) **51**; English translation, with historical comments: Gen. Rel. Grav. **29** 637 (1997)
3. R.C. Tolman, Proc. Nat. Acad. Sci. USA **20** (1934) 169; reprinted, with historical comments: Gen. Rel. Grav. **29** 931 (1997)
4. H. Alnes, M. Amarzguioui, O. Gron, Phys. Rev., **D73** 08351 (2006)
5. J. García-Bellido, T. Haugbolle, J. Cosmol Astropart. Phys. **04** 003 (2008)
6. K. Bolejko and J. S. B. Wyithe, J. Cosmol. Astropart. Phys. 02 (2009) 020.
7. H. Alnes, M. Amarzguioui Phys. Rev **D75** 023506 (2007)
8. H. Alnes, M. Amarzguioui, Phys. Rev. D75 (2007) 023506.
9. K. Bolejko and M.-N. Célérier, Phys. Rev. **D82** 103510 (2010)
10. K.T. Inoue and J. Silk, Astrophys. J. **648** 23 (2006)
11. K.T. Inoue and J. Silk, Astrophys. J. **664** 650 (2007)
12. K. Bolejko and R.A. Sussman, Phys Lett B **697** 265–270
13. P. Szekeres, Commun. Math. Phys. **41** 55 (1975)
14. C. Hellaby, A. Krasiński, Phys. Rev. **D77** 023529 (2008)
15. A. Krasiński Phys. Rev, **D78** 064038 (2008)
16. C. Hellaby, A. Krasiński, Phys. Rev. **D66** 084011 (2002)
17. K.L. Thompson and E.T. Vishniac, Astrophys. J. **313** 517 (1987)
18. K. Tomita, Astrophys. J. **529** 38 (2000)
19. C. Clarkson, M. Regis, arXiv:1007.3443 (2010)
20. P.J.E. Peebles, The Large-Scale Structure of the Universe. Princton University Press, Princton (1980)
21. K. Bolejko, Phys. Rev. **D73** 123508 (2006)
22. K. Bolejko, Phys. Rev. **D75** 043508 (2007)
23. K. Bolejko, Gen. Rel. Grav. **41** 1585 (2009)
24. R. van de Weygaert and W. Schaap in Data Analysis in Cosmology, ed. V. Martínez, E. Saar, E. Martínez-González, M. Pons-Bordería, Springer-Verlag, Berlin, Lecture Notes in Physics **665** p. 291 (2009)

[3] http://msowww.anu.edu.au/skymapper/

25. A. Dekel, et al., Astrophys. J. **522** 1 (1999)
26. J. Plebański, A. Krasiński, An introduction to general relativity and cosmology. Cambridge University Press, Cambridge (2006)
27. T. Buchert, Gen. Rel. Grav. **40** 467 (2008); R. Zalaletdinov, Int. J. Mod. Phys. **A23** 1173 (2008)
28. M. Ishak, J. Richardson, D. Garred, D. Whittington, A. Nwankwo, R. Sussman, Phys. Rev. **D78** 123531 (2008)
29. A. Nwankwo, J. Thompson, M. Ishak, arXiv1005.2989 (2010)
30. A. Krasiński, K. Bolejko, arXiv:1007.2083 (2010)

Chapter 12
Finding New Signature Effects on Galactic Dynamics to Constrain Bose–Einstein-Condensed Cold Dark Matter

Tanja Rindler-Daller and Paul R. Shapiro

Abstract If cosmological cold dark matter (CDM) consists of light enough bosonic particles that their phase–space density exceeds unity, they will comprise a Bose–Einstein condensate (BEC). The nature of this BEC-CDM as a quantum fluid may then distinguish it dynamically from the standard form of CDM involving a collisionless gas of non-relativistic particles that interact purely gravitationally. We summarize some of the dynamical properties of BEC-CDM that may lead to observable signatures in galactic halos and present some of the bounds on particle mass and self-interaction coupling strength that result from a comparison with observed galaxies.

12.1 Introduction

Astronomical observations suggest the presence of non-baryonic, non-relativistic ('cold') dark matter (DM), comprising around 23 % of the energy density in the Universe. The particle nature of dark matter remains elusive, however, despite ongoing efforts to detect it directly in search experiments or indirectly via imprints on astrophysical observations. In large N-body simulations of structure formation, cold dark matter (CDM) has been modeled as a collisionless gas, which only interacts gravitationally. We refer to this as standard CDM. Despite the successes in reproducing the large-scale structure, as well as reproducing flat rotation curves at the outskirts of galaxies, standard CDM seems to be in conflict with observations of galactic small-scale properties. These are notably the overabundance of subhalos around big hosts of Milky-Way size and beyond, and the failure to reproduce flat cores in the centers of dark-matter dominated dwarf and LSB galaxies. Both features apparently contradict astronomical observations, and have been the subject of active research in the past decade. One approach has been to study the combined effects

T. Rindler-Daller (✉) · P. R. Shapiro
Department of Astronomy and Texas Cosmology Center,
The University of Texas at Austin, 2515 Speedway C1400, Austin TX, USA
e-mail: daller@astro.as.utexas.edu

P. R. Shapiro
e-mail: shapiro@astro.as.utexas.edu

C. Moreno González et al. (eds.), *Accelerated Cosmic Expansion*,
Astrophysics and Space Science Proceedings 38,
DOI 10.1007/978-3-319-02063-1_12,

of the collisionless CDM component and the dissipative baryonic component, to see if the dissipative hydrodynamics of the latter can affect a cure. Another possible solution has been to go beyond the simplifying assumption that the DM particles are cold and/or collisionless. As a result, there has been a recent revival of investigations of structure formation with non-standard dark matter candidates, like warm dark matter e.g. Schneider et al.[29], self-interacting fermionic dark matter (SIDM) e.g. Ahn & Shapiro [1], Koda & Shapiro [19] and Bose–Einstein-condensed dark matter (BEC-CDM or scalar-field dark matter), e.g. Woo & Chiueh [39] and Suárez & Matos [35], all proposals of which are able to suppress the formation of self-bound structures below a certain scale, as well as to flatten central profiles, depending on the respective particle parameters.

In this paper, we will address BEC-CDM. We are particularly interested in a class of models describing self-interacting bosonic dark matter, the particles of which are so light that they collectively occupy their ground state below a certain temperature, forming a Bose–Einstein condensate (BEC) in the early Universe. Then, this state can be well described by a scalar field, the so-called wavefunction of the condensate. We will assume that the wavefunction respects a $U(1)$-symmetry, such that the number of particles is conserved. There is thus no self-annihilation of DM in this scenario, in contrast to models considered for instance by Tkachev [36]. Recent advances in particle theory predict the generic existence of bosons as light as or much lighter than the QCD axion—the 'classic' bosonic DM candidate, values ranging from about 10^{-33} to $\gtrsim 10^{-5}$ eV/c^2, which can serve as the CDM in the Universe (see e.g. Günther & Zhuk [15], Carroll [8], Arvanitaki et al. [4]). The theoretical description of these very light bosons in terms of scalar fields leads to halo dynamics which can be understood as the solution of nonlinear wave equations (e.g. Alcubierre et al.[2], Chavanis [10], Matos & Ureña-López [24], Sin [33], Ureña-López & Guzmán [37]), in contrast to the N-body dynamics of standard CDM. As a matter of fact, much less investigation has been pursued in the literature so far, in particular with respect to the nonlinear stages of structure formation for this type of DM. Therefore, we are in need of a better understanding of how much of the parameter space of this form of dark matter is able to reproduce the successes, while resolving the failures, of standard CDM.

In what follows, we will summarize some of the recent work in this field, highlighting some of our own, with apologies for the fact that length limitations prevent us from attempting a more comprehensive review. In Sect. 12.2, we will describe the basic equations which govern BEC-CDM dynamics, showing how its quantum nature leads to fluid behavior. We will then distinguish two regimes according to the strength of the particle self-interaction, and state how the characteristic size and mass of structures that form gravitationally from this form of DM are related to the particle mass m and self-interaction coupling strength g. We also present the virial theorem for isolated BEC-CDM halos. In Sect. 12.3, we summarize how the equilibrium structure of BEC-CDM halos can distinguish them from halos in standard CDM, for halos with and without rotation. We show that halos in the BEC-CDM model typically rotate with enough angular momentum that, if they are of the type supported against gravitational collapse by self-interaction pressure and rotation, quantum vortices are likely to form, which can affect halo density profiles. We also summarize a few examples in which BEC-CDM can be distinguished by its effect on baryonic structures. In Sect. 12.4, we summarize the bounds on particle mass and coupling strength that follow from the requirement that halos or halo cores not exceed some characteristic size, if halos or their cores are treated as isolated, equilibrium structures. Finally, in Sect. 12.5, we argue that, for BEC-CDM to reproduce the full range of structure scales found in our Universe (as in standard CDM), we must go beyond the modeling of individual objects in static equilibria to account for continuous mass infall and its dynamical consequences.

12.2 Quantum-Coherent Dark Matter Under Newtonian Gravity

12.2.1 Fundamental Properties

The dynamical description of dark matter within galactic halos usually involves small gravitational and velocity fields. With regard to BEC-CDM, it has proved advantageous to consider the non-relativistic Schrödinger–Poisson, or Gross–Pitaevskii–Poisson (GPP) system of equations of motion for the dark matter BEC wavefunction ψ, as follows

$$i\hbar\frac{\partial\psi}{\partial t} = -\frac{\hbar^2}{2\,m}\Delta\psi + m\Phi\psi + g|\psi|^2\psi, \tag{12.1}$$

$$\Delta\Phi = 4\pi Gm|\psi|^2. \tag{12.2}$$

The terms on the rhs in (12.1), which govern the evolution, are due to the quantum-kinetic energy, the gravitational potential Φ, and the self-interaction of identical bosons. The latter has been included in the usual way in terms of an effective interaction potential $g|\psi|^4/2$ with coupling constant (or self-interaction strength) g. The possibly complicated particle interactions are simplified this way in the low-energy limit of a dilute gas: disregarding higher than two-body interactions, the cross section for elastic scattering of indistinguishable bosons becomes constant in the low-energy limit,

$$\sigma = 8\pi a_s^2 \tag{12.3}$$

with the s-wave scattering length a_s. The coupling constant of the effective interaction is then given by

$$g = 4\pi\hbar^2 a_s/m. \tag{12.4}$$

We shall note that the above GP equation is strictly valid only for dilute systems, which means that a_s must be much smaller than the mean interparticle distance, i.e. $a_s \ll n^{-1/3}$. Also, we are restricted to $g \geq 0$ ($a_s \geq 0$), because models described by (12.1) with negative self-interaction coupling g are not able to provide stable structures, since the resulting negative pressure works in favor of gravity. If we assume that all of the DM is in the condensed state, the number and mass density of DM in the halo is given by $n(\mathbf{r}) = |\psi|^2(\mathbf{r})$ and $\rho(\mathbf{r}) = mn(\mathbf{r})$, respectively.

The equations can be written in fluid-like form by inserting into (12.1) the polar decomposition of the wavefunction (applied early by Madelung [23] to the free-particle Schrödinger equation),

$$\psi(\mathbf{r},t) = |\psi|(\mathbf{r},t)e^{iS(\mathbf{r},t)} = \sqrt{\frac{\rho(\mathbf{r},t)}{m}}e^{iS(\mathbf{r},t)}, \tag{12.5}$$

resulting in the momentum and continuity equations,

$$\rho\frac{\partial \mathbf{v}}{\partial t} + \rho(\mathbf{v}\cdot\nabla)\mathbf{v} = -\rho\nabla Q - \rho\nabla\Phi - \nabla P_{SI}, \tag{12.6}$$

$$\frac{\partial \rho}{\partial t} + \nabla\cdot(\rho\mathbf{v}) = 0, \tag{12.7}$$

with the bulk velocity $\mathbf{v} = \hbar\nabla S/m$. The gradient of

$$Q = -\hbar^2\Delta\sqrt{\rho}/(2\,m^2\sqrt{\rho}) \tag{12.8}$$

gives rise to what is often called 'quantum pressure', an additional force on the rhs of Eq. (12.6), which basically stems from the quantum-mechanical uncertainty principle. The particle self-interaction, on the other hand, gives rise to a pressure of polytropic form

$$P_{SI} = g\rho^2/(2\,m^2). \tag{12.9}$$

The quantum-kinetic term provides an important characteristic length scale, as follows: with Δ having a dimension of L^{-2} and changing to the momentum representation, one can easily see that the characteristic length is essentially nothing but the de Broglie length of the bosons

$$L \sim \lambda_{deB} = h/p = h/(mv). \tag{12.10}$$

Since we will consider virialized, isolated objects, it makes sense to use the corresponding virial velocity in this expression for λ_{deB}, assuming that the particles stay in their Bose–Einstein-condensed state after virialization, which is the case for the model parameters we are going to encounter in this paper (see also Sect. 12.4). BEC-CDM without self-interaction, $P_{SI} = 0$, was termed 'Fuzzy Dark Matter' by Hu, Barkana & Gruzinov [16]. In this review, we will call it BEC-CDM of TYPE I. In this regime, it is the quantum-kinetic term via (12.10) which determines the

Table 12.1 Lower bound on the boson mass, provided by the 'Fuzzy Dark Matter' regime, or TYPE I BEC-CDM, for different cosmological structures

	Halo mass [$M_\odot$]	Size [kpc]	Boson mass [eV]
Milky Way (MW)	10^{12}	100	$1.066 \cdot 10^{-25}$
Dwarf galaxy (DG)	10^{10}	10	$3.371 \cdot 10^{-24}$
Dwarf spheroidal (dSph)	10^{8}	1	$1.066 \cdot 10^{-22}$
Minihalo (MH)	10^{6}	0.1	$3.371 \cdot 10^{-21}$

equilibrium size of self-gravitating objects - in order for λ_{deB} not to exceed a certain galactic length scale, the particle mass m must be sufficiently large. Table 12.1 contains different typical halo sizes and the corresponding lower limits on m.

However, it has been observed in the previous literature, e.g. Colpi, Shapiro & Wasserman [12] and Lee & Lim [21], and we confirm it as well, see Eq. (12.13) and Sect. 12.3, that a larger mass m than inferred from (12.10) can result in stable structures of the same given size, if self-interaction is included. In fact, considering the case in which the last term in (12.6) supports the system against gravitational collapse, while $Q = 0$, we arrive at the opposite regime to TYPE I, which we call the Thomas–Fermi regime of BEC-CDM, or TYPE II for short. We note that this regime goes under many names; Goodman [13] calls it 'repulsive dark matter' (RDM), Peebles [26] speaks of 'fluid dark matter'. We also note that both regimes have already been considered for related models by Khlopov, Malomed & Zeldovich [18], in studying gravitational instabilities of a primordially produced scalar field.

In the following, we will make use of convenient units, defined as in Rindler–Daller & Shapiro (RDS) [28]:

$$m_H \equiv \frac{\hbar}{R^2(\pi G \bar{\rho})^{1/2}} = \frac{2\hbar}{\sqrt{3G}}(RM)^{-1/2}$$

$$= 1.066 \cdot 10^{-22} \left(\frac{R}{1\ \mathrm{kpc}}\right)^{-1/2} \left(\frac{M}{10^8\ M_\odot}\right)^{-1/2} \mathrm{eV}, \qquad (12.11)$$

and

$$g_H \equiv \hbar^2/(2\bar{\rho}R^2) = 2\pi\hbar^2 R/(3M)$$

$$= 2.252 \cdot 10^{-62} \left(\frac{R}{1\ \mathrm{kpc}}\right) \left(\frac{M}{10^8\ M_\odot}\right)^{-1} \mathrm{eV\ cm}^3, \qquad (12.12)$$

with $c = 1$. We described the meaning and significance of those parameters at length in RDS [28]. It shall be sufficient to re-iterate here that m_H is the characteristic mass of a non-interacting particle whose de Broglie wavelength is comparable to the size of a given halo, see Table 12.1. It is thus the smallest particle mass possible in order for quantum pressure to be solely responsible for holding that halo up against gravitational collapse. On the other hand, if there are density variations in the BEC fluid of scale length R, then g_H is the coupling strength for which the quantum and self-interaction pressure force terms are equal.

It turns out that the TYPE II regime is a good approximation as long as $g/g_H \gg 2$ is fulfilled, as we demonstrated in [28]: to determine whether a BEC-CDM halo of a given size R is of TYPE I or TYPE II, we have to compare the quantum pressure and self-interaction pressure terms in (12.6) to each other,

$$|-\rho\nabla Q|/|-\nabla P_{SI}| \sim \hbar^2/(g\rho R^2) \sim 2g_H/g \ll 1,$$

from which the claim follows. We have shown in RDS [28], Eq. (46), that the radius of a spherical halo is then related to the de Broglie length of the boson according to

$$R_0 = \frac{\sqrt{3}\pi^{1/4}}{12}\left(\frac{g}{g_H}\right)^{1/2} \lambda_{deB}. \tag{12.13}$$

Hence, since $g/g_H \gg 2$, $R_0 \gg \lambda_{deB}$ for TYPE II BEC-CDM halos.

12.2.2 Stationary Systems and Virial Equilibrium

In the context of BEC-CDM in the GPP framework, Eq. (12.1) and (12.2), stationary self-gravitating halos can be described by wavefunctions of the form

$$\psi(\mathbf{r}, t) = \psi_s(\mathbf{r})e^{-i\mu t/\hbar}, \tag{12.14}$$

where the conservation of particle number fixes μ, the GP chemical potential. While ψ evolves harmonically in time, the mass density $\rho = m|\psi_s|^2$ and, hence, the gravitational potential Φ are time-independent. Inserting this ψ into (12.1) results in the time-independent GP equation with eigenvalues μ,

$$\left(-\frac{\hbar^2}{2m}\Delta + g|\psi_s|^2 + m\Phi\right)\psi_s = \mu\psi_s. \tag{12.15}$$

The time-independent part $\psi_s(\mathbf{r})$ itself can be decomposed as usual,

$$\psi_s(\mathbf{r}) = |\psi_s|(\mathbf{r})e^{iS_s(\mathbf{r})} \tag{12.16}$$

with both amplitude and phase depending here on position only. We will omit the subscript 's' in the forthcoming analysis. Systems obeying (12.15) can be studied via the corresponding GP energy functional given by

$$\mathcal{E}[\psi] = \int_V \left[\frac{\hbar^2}{2m}|\nabla\psi|^2 + \frac{m}{2}\Phi|\psi|^2 + \frac{g}{2}|\psi|^4\right] d^3\mathbf{r}. \tag{12.17}$$

Inserting (12.16) into (12.17), the total energy can be written as

$$E = K + W + U_{SI}, \tag{12.18}$$

with the total kinetic energy term

$$K \equiv \int_V \frac{\hbar^2}{2\,m}|\nabla\psi|^2 d^3\mathbf{r} = \int_V \frac{\hbar^2}{2\,m^2}(\nabla\sqrt{\rho})^2 d^3\mathbf{r} + \int_V \frac{\rho}{2}\mathbf{v}^2 d^3\mathbf{r} \equiv K_Q + T. \tag{12.19}$$

K_Q accounts for the quantum-kinetic energy and T for the bulk kinetic energy of the body, which comes in the form of rotation or internal motion. K_Q has no classical counterpart, and is absent in the classical figures of equilibrium studied in the previous literature. Also, K_Q is neglected in the TYPE II regime. The other terms in (12.18) are the gravitational potential energy

$$W \equiv \int_V \frac{\rho}{2}\Phi d^3\mathbf{r} \tag{12.20}$$

and the internal energy

$$U_{SI} \equiv \int_V \frac{g}{2\,m^2}\rho^2 d^3\mathbf{r}, \tag{12.21}$$

which is determined by the particle interactions, and which we have defined essentially as $U_{SI} = \int P_{SI} dV$ with P_{SI} in (12.9). The origin of U_{SI} is due to the repulsive two-body elastic scattering of identical bosons, Eq. (12.3). The above energy contributions enter the scalar virial theorem of an *isolated* (possibly rotating) BEC halo under self-gravity, which reads as

$$2\,K + W + 3U_{SI} = 0. \tag{12.22}$$

As in classical gas dynamics, (12.22) can be derived by multiplying the equations of motion in fluid form, Eq. (12.6), by $\mathbf{r}$ and integrating the resulting equation over volumes which enclose the system of interest. For an isolated body, a derivation involving a scaling argument was presented by Wang [38].

12.3 Signature Effects of BEC-CDM on Halos and Halo Cores

12.3.1 Sizes and Density Profiles

The equilibrium density profiles of self-gravitating BEC-CDM halos are solutions of (12.1) or (12.6)–(12.9) respectively, along with (12.2). As an important result, they are universal in shape. Furthermore, BEC-CDM halos have a finite central density. In fact, this last feature has been one of the motivations in the previous literature to consider this form of DM as a solution to the cusp–core problem of dark matter dominated dwarf galaxies (DG) and dwarf-spheroidal (dSph) galaxies.

In the case of TYPE I, the density profile can only be determined numerically: it falls off as r^{-4} for large r, but has no compact support. The radius which includes 99 % of the mass reads as

$$R_{99} = 9.9\hbar^2/(GMm^2) \tag{12.23}$$

(see Membrado et al. [25] for more details). For TYPE II, on the other hand, the equation of state reduces to an $(n = 1)$-polytrope, with P_{SI} in (12.9) and $Q = 0$, having the well-known spherical density profile

$$\rho^S(r) = \rho_c^S sinc(\sqrt{4\pi Gm^2/g}\ r), \tag{12.24}$$

with $sinc(x) \equiv sin(x)/x$ and the central density ρ_c^S. The corresponding halo radius is then given by

$$R_0 = \pi\sqrt{\frac{g}{4\pi Gm^2}}, \tag{12.25}$$

see e.g. Goodman [13] for more details.

In both regimes, TYPE I and TYPE II, the halo profile and size are determined by the DM particle parameters. Of course, this is also true for the intermediate regime: Chavanis and Delfini [11] calculate numerical solutions for the mass–radius relationship, $R = R(M)$, which interpolate between TYPE I and II. However, for any given particle model, neither (12.23) and (12.25) nor the result of [11] are able to re-produce the fact that R must increase with M as we know from astronomical observations. The successful fitting of galaxy data using the associated rotation curves of TYPE I and TYPE II BEC-CDM halos by Arbey et al. [3] and Böhmer and Harko [7] must thus be judged with this caveat in mind. It implies that it is necessary to go beyond the description of non-rotating halos in virial equilibrium, composed of a pure BEC-CDM fluid, if this DM model is to describe galactic structures successfully.

12.3.2 Rotation and Shape

In RDS [28], we studied the properties of BEC-CDM halos in the TYPE II case, once rotation is taken into account. It is generally believed that tidal torques caused by large-scale structure give a halo most of its angular momentum in the early phases of halo collapse. This picture has been confirmed by cosmological N-body simulations of the standard CDM universe, which show that halos form with a net angular momentum such that the dimensionless ratio, the so-called spin parameter,

$$\lambda = \frac{L|E|^{1/2}}{GM^{5/2}}, \tag{12.26}$$

where L is the total angular momentum and E the total energy of the halo, has typical values in the range of $[0.01, 0.1]$ with a median value $\simeq 0.05$ (see e.g. Barnes and

Efstathiou [5]). The degree of rotational support is thus very small for the CDM halos which surround galaxies. We will be interested in the case where the BEC nature of DM affects small-scale structure and the internal dynamics of halos, while large-scale structure formation shall follow the ΛCDM model to a great extent. Therefore, we adopt the above range of spin parameters for BEC-CDM halos, too.

We describe the effect of rotation on the structure of BEC-CDM halos by two approximations which are based upon the classic models of rotating figures of equilibrium (see e.g. Chandrasekhar [9]). The simplest description assumes that the halos are Maclaurin spheroids, which are axisymmetric, oblate, and homogeneous. Not only is this model fully analytical, it also provides a convenient background solution to perturb in determining if and when quantum vortex formation is energetically favored. In the absence of quantum vortices, however, BEC-CDM is *irrotational*, while the Maclaurin spheroid model assumes uniform rotation. Of course, this irrotationality can be broken locally in the fluid, by creating quantum vortices, if the amount of angular momentum exceeds a certain minimum, as shown below. In the limit of large enough angular momentum that a vortex *lattice* develops, in fact, uniform rotation is a good approximation, and so will the Maclaurin spheroids be. More generally, to account for irrotationality, we also consider a second model, that of irrotational Riemann-S ellipsoids. Since the classic solution for Riemann-S ellipsoids is homogeneous, however, we account for the $(n = 1)$-polytropic nature of TYPE II BEC-CDM by adopting the solution derived by Lai, Rasio & Shapiro (LRS) [20] for *compressible* Riemann-S ellipsoids, based on the 'ellipsoidal approximation' which assumes self-similar ellipsoidal density strata.

We denote the semi-axes of those bodies along (x, y, z) as (a_1, a_2, a_3). Maclaurin spheroids fulfill $a_1 = a_2 > a_3$ with eccentricity $e = (1-(a_3/a_1)^2)^{1/2}$. Using (12.22), we can determine how the (mean) radius $R = (a_1a_2a_3)^{1/3}$ and the spin parameter of such a halo depend on its eccentricity, see RDS [28]:

$$R = \left(\frac{15}{3A_3(e)(1-e^2)^{2/3}}\right)^{1/2} \left(\frac{g}{4\pi G m^2}\right)^{1/2}, \tag{12.27}$$

$$\lambda = \frac{6}{5\sqrt{5}} \frac{\arcsin e}{e} t \left(1 + \frac{e}{t} \frac{A_3(e)(1-e^2)^{1/2}}{\arcsin(e)}\right)^{1/2}, \tag{12.28}$$

with the t-parameter $t \equiv T/|W|$, a measure of rotational support, given by (see also LRS [20])

$$t(e) = 3/(2e^2) - 1 - 3\sqrt{1-e^2}/(2e \arcsin(e)), \tag{12.29}$$

and $A_3(e) = 2/e^2 - 2\sqrt{1-e^2}\arcsin(e)/e^3$ for this model. On the other hand, a compressible, irrotational Riemann-S ellipsoid of polytropic index $n = 1$ must be prolate, i.e. its semi-axes fulfill $a_1 \geq a_3 \geq a_2$, and the eccentricities are given by $e_1 = (1-(a_2/a_1)^2)^{1/2}$ and $e_2 = (1-(a_3/a_1)^2)^{1/2}$. In that case, the expressions for the mean radius and spin parameter

$$R = R_0 g(e_1, e_2)^{-1/2}, \quad \lambda = \lambda(e_1, e_2) \tag{12.30}$$

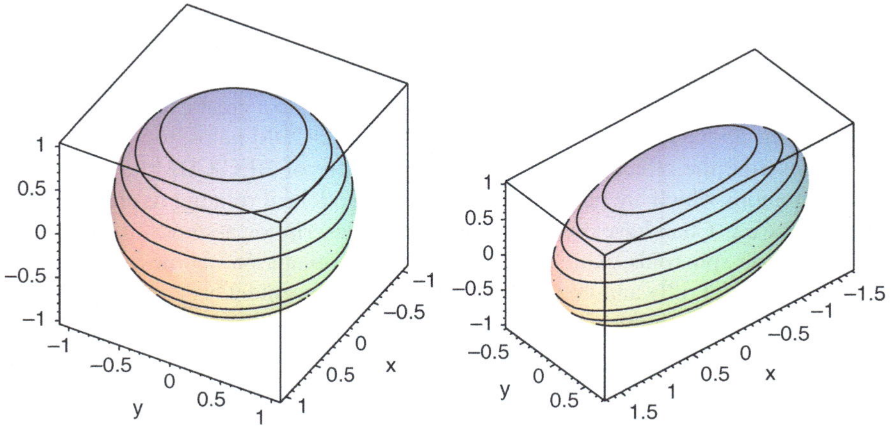

Fig. 12.1 Halos rotating about the z-axis for $a_3 = 1$ and $\lambda = 0.05$: Maclaurin spheroid having $e = 0.302$ (*left-hand-plot*) and irrotational Riemann-S ellipsoid having $(e_1, e_2) = (0.881; 0.797)$ (*right-hand-plot*); see also RDS [28]

with R_0 in (12.25) and $g(e_1, e_2)$ and $\lambda(e_1, e_2)$ functions of the eccentricities, are very cumbersome and we refer to RDS [28], Eq. (101) for more details.

For both models, the characteristic size—the mean radius—depends on the particle parameters exactly the same way as in the non-rotating case, namely $R \sim \sqrt{g}/m$. The effect of the rotation is thus only to change the overall multiplicative factor of this dependence as seen from (12.27) and (12.30). By fixing $\lambda = (0.01, 0.05, 0.1)$, we can solve $\lambda = \lambda(e)$ in (12.28) and $\lambda = \lambda(e_1, e_2)$ in (12.30), respectively, for the eccentricities. Figure 12.1 shows two illustrative examples of our rotating halo models.

We were able in RDS [28] to derive the generalization to Eq. (12.24) for the case of the $(n = 1)$-polytropic Riemann-S ellipsoid analytically, assuming the ellipsoidal approximation of LRS [20]. Accordingly, the ellipsoidal density profile reads

$$\rho^E(\tilde{q}) = \rho_c^E sinc\left[\tilde{q}(1 - e_1^2)^{1/6}(1 - e_2^2)^{1/6} g(e_1, e_2)^{1/2}\right] \qquad (12.31)$$

with

$$\tilde{q}^2 = [x^2 + y^2/(1 - e_1^2) + z^2/(1 - e_2^2)](\pi/R_0)^2,$$

ρ_c^E the central density of the ellipsoid, and R_0 in (12.25). It is plotted for different λ in Fig. 12.2.

The virial theorem (12.22) not only provides the above relationships for the mean radius and spin-parameter of a halo, but also relates the DM particle parameters in a characteristic way: it can be shown that the corresponding formulae are

$$\frac{m}{m_H} = \left(\frac{5}{8A_3(e)(1 - e^2)^{2/3}}\right)^{1/2} \sqrt{\frac{g}{g_H}} \qquad (12.32)$$

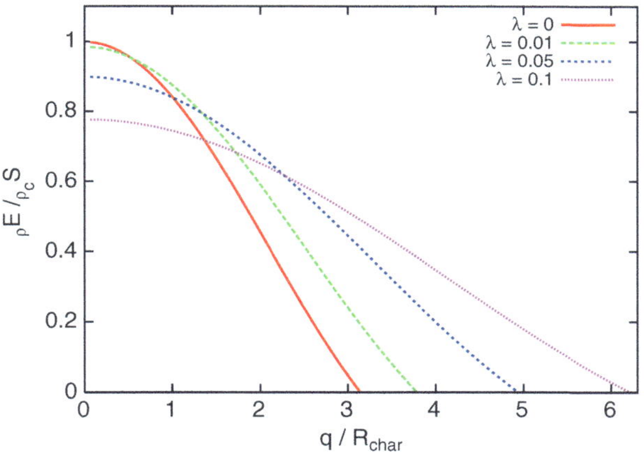

Fig. 12.2 Density profiles of the (vortex-free) ($n = 1$)-polytropic Riemann-S ellipsoidal halos having $\lambda = (0.01, 0.05, 0.1)$, according to Eq. (12.31). The profile of the spherical halo, Eq. (12.24), is added for comparison (solid curve). The densities are all normalized to $\rho_c^S = 1$; $q = \tilde{q} R_{char} \equiv \tilde{q}[g/(4\pi G m^2)]^{1/2}$. The locus of the outer surface, where the density vanishes, increases with λ; see also RDS [28]

for Maclaurin spheroidal halos, and

$$\frac{m}{m_H} = \frac{\pi}{\sqrt{8}} g(e_1, e_2)^{-1/2} \sqrt{\frac{g}{g_H}} \tag{12.33}$$

for Riemann-S ellipsoidal halos, see RDS [28]. That is, virialized rotating halos, according to either model, will lie on a straight line in $(\log m, \log g)$-space, whose slope is completely determined by the eccentricities or mean radius of a given halo. These relationships can be found in Figs. 12.4 and 12.5, respectively. From those and the condition $g/g_H \gg 2$, it follows that, in the TYPE II regime, $m/m_H \gg 1$, as it must be for $\lambda_{deB} \ll R_0$.

In [28], we also determined the conditions for the formation of a central quantum vortex in a given rotating halo. The minimum amount of angular momentum *necessary* to form a singly-quantized, axisymmetric vortex in the center of a halo with total number of particles $N = M/m$ is given by

$$L_{QM} \equiv N\hbar, \tag{12.34}$$

which, for a given λ and L becomes a minimum condition on the particle mass. The relationship is linear for both halo models, i.e.

$$m/m_H = f(e_1, e_2) L/L_{QM}, \tag{12.35}$$

with $f(e_1, e_2)$ denoting here another function depending only on the eccentricities (see Eqs. (108) and (109) in [28]). In order to find a criterion which is, not only *necessary*, but also *sufficient*, we determined the DM particle parameters for which

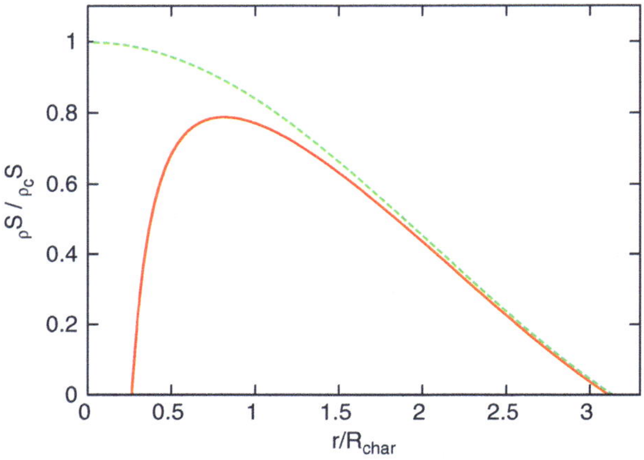

Fig. 12.3 Density profile of a spherical halo with vortex in the center (solid line) which transitioned from a Riemann-S ellipsoidal halo having $\lambda = 0.05$. The profile of the spherical halo with $L = 0$ and no vortex (dashed line) is added for comparison; R_{char} as in Fig. 12.2. See also RDS [28]

the central vortex is energetically favored, i.e. the conditions which make the halo with vortex have less energy than a rotating, but otherwise vortex-free halo. For this purpose, we consider two limiting cases: *Model A* describes a halo with a high enough angular momentum that $L \gg L_{QM}$, i.e. the central vortex is essentially considered to be only a small perturbation of the total angular momentum L of the halo. For this case, we use Maclaurin spheroids as our halo model. *Model B*, on the other hand, assumes a given halo has just enough angular momentum to form one quantum vortex, i.e. $L = L_{QM}$. We use the irrotational, $(n = 1)$-polytropic Riemann-S ellipsoids for this case. In this case, once the central vortex is energetically favored, its formation takes up all of the angular momentum, leaving a spherical halo with vortex, see Fig. 12.3 for an example.

Despite differences in the assumed halo equilibrium models, the conclusions are the same for *Model A* and *Model B*: vortex formation requires a minimum particle mass $m \geq m_{crit}$ and a minimum particle self-interaction coupling strength $g \geq g_{crit}$. For any (m, g) pair which satisfies $m \geq m_{crit}$ and $g \geq g_{crit}$ at a given λ, this pair will also favor vortex formation for any *larger* value of λ. Our main results are summarized in Figs. 12.4 and 12.5, which are independent of halo size, respectively, while Tables 12.2 and 12.3 list physical units for m_{crit}, g_{crit} and the corresponding vortex size, for the cases of a typical dwarf galaxy (DG) and a dwarf spheroidal galaxy (dSph), according to Table 12.1. Since we fix λ and $L = L_{QM}$ for the case of *Model B*, the respective particle parameters $((m/m_H)_{crit}, (g/g_H)_{crit})$ are uniquely determined via Eqs. (12.35) and (12.33), respectively (see also Fig. 12.5 for those numbers). It turns out that for all λ considered, vortex formation is favored for those parameters. We interpret this to mean that, for these same λ-values, if $L/L_{QM} > 1$, instead, (i.e. $m/m_H > (m/m_H)_{crit}$, according to (12.35)), vortex formation will also be favored, as long as $g/g_H > (g/g_H)_{crit}$. On the other hand, for halos with $L < L_{QM}$, the particle parameters must satisfy $m/m_H < (m/m_H)_{crit}$ and $g/g_H < (g/g_H)_{crit}$, respectively, so they will *not* form vortices, but can still be modeled as Riemann-S ellipsoids. (This assumes, of course, that $g/g_H \gg 2$, so they are still in the TYPE II regime).

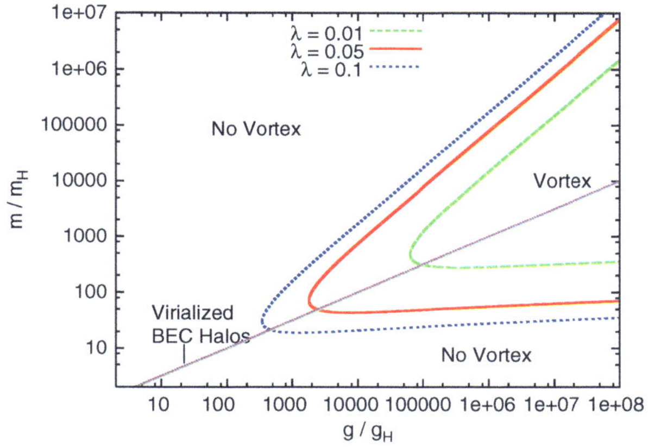

Fig. 12.4 *Model A:* Curves within which vortex formation is energetically favored in the dimensionless BEC-CDM particle parameter space $(m/m_H, g/g_H)$ for halos with spin parameter $\lambda = 0.01, 0.05, 0.1$, respectively. Straight lines for the same λ- values: halos fulfilling virial equilibrium, according to Eq. (12.32); see also RDS [28]

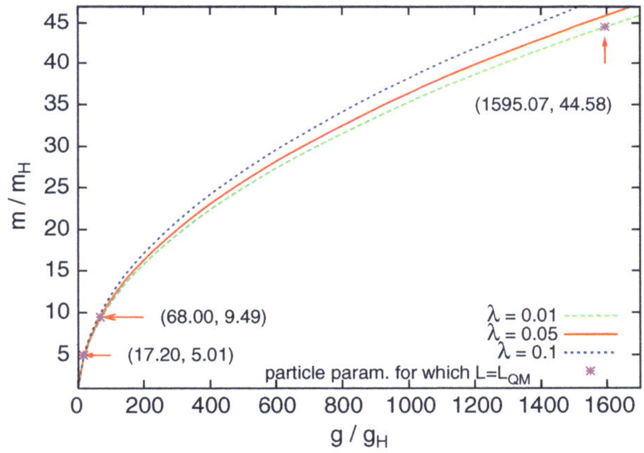

Fig. 12.5 *Model B:* Dimensionless BEC-CDM particle parameter space $(m/m_H, g/g_H)$ (no log-scale!) for halos with spin parameter $\lambda = 0.01, 0.05, 0.1$, respectively, fulfilling virial equilibrium, according to Eq. (12.33). For each λ, dots denote those critical particle parameters for which $L = L_{QM}$; see also RDS [28]

Table 12.2 Lower bounds on the boson mass and self-interaction coupling strength for vortex formation in TYPE II BEC-CDM halos with a given spin parameter λ in *Model A*. ξ_{max}-values denote the upper bounds for the corresponding vortex core radius

Dwarf galaxy			
λ	m_{crit} [eV]	g_{crit} [eV cm^3]	ξ_{max} [kpc]
0.01	$1.04 \cdot 10^{-21}$	$2.30 \cdot 10^{-58}$	0.03
0.05	$1.67 \cdot 10^{-22}$	$5.74 \cdot 10^{-60}$	0.20
0.10	$7.33 \cdot 10^{-23}$	$1.02 \cdot 10^{-60}$	0.47
Dwarf spheroidal galaxy			
λ	m_{crit} [eV]	g_{crit} [eV cm^3]	ξ_{max} [kpc]
0.01	$3.30 \cdot 10^{-20}$	$2.30 \cdot 10^{-57}$	$3.13 \cdot 10^{-3}$
0.05	$5.28 \cdot 10^{-21}$	$5.74 \cdot 10^{-59}$	0.02
0.10	$2.32 \cdot 10^{-21}$	$1.02 \cdot 10^{-59}$	0.05

Table 12.3 Lower bounds on the boson mass and self-interaction coupling strength for vortex formation in TYPE II BEC-CDM halos with a given spin parameter λ in *Model B*. ξ_{max}-values denote the upper bounds for the corresponding vortex core radius

Dwarf galaxy			
λ	m_{crit} [eV]	g_{crit} [eV cm^3]	ξ_{max} [kpc]
0.01	$1.50 \cdot 10^{-22}$	$3.59 \cdot 10^{-60}$	0.25
0.05	$3.20 \cdot 10^{-23}$	$1.53 \cdot 10^{-61}$	1.21
0.10	$1.69 \cdot 10^{-23}$	$3.87 \cdot 10^{-62}$	2.41
Dwarf spheroidal galaxy			
λ	m_{crit} [eV]	g_{crit} [eV cm^3]	ξ_{max} [kpc]
0.01	$4.75 \cdot 10^{-21}$	$3.59 \cdot 10^{-59}$	0.02
0.05	$1.01 \cdot 10^{-21}$	$1.53 \cdot 10^{-60}$	0.12
0.10	$5.34 \cdot 10^{-22}$	$3.87 \cdot 10^{-61}$	0.24

We shall comment on other previous work on vortices in related models. Silverman & Mallet [32] give heuristic arguments for vortex formation, but do not derive the critical conditions or their consequences for the DM particle parameters. Yu & Morgan [40] show that vortex lattices can provide flat rotation curves for galaxies. Kain & Ling [17], on the other hand, find approximate solutions for the density profile of a nonrotating, spherically-symmetric halo with a single vortex that contains all the angular momentum for the TYPE II case (as in Fig. 12.3, solid curve). Their estimates for the viable parameter space of particle mass for the Andromeda galaxy is in agreement with our more precise results for the Milky Way, see RDS [28].

Our results strongly suggest that vortices can only form in BEC-CDM with a sufficiently large, positive self-interaction coupling strength. Vortex formation in axion DM *without self-interaction*, which has been claimed by Sikivie & Yang [31], seems thus not viable according to our results. However, for TYPE II halos, i.e. those for which $m/m_H \gg 1$ and $g/g_H \gg 2$, vortex formation is favored for the range of halo spin parameters found in the CDM model, for a large portion of the BEC particle parameter space. For example, for $\lambda = 0.05$, *Model B* yields $(m/m_H)_{crit} = 9.49$ and $(g/g_H)_{crit} = 68.00$, while *Model A* yields $(m/m_H)_{crit} = 49.52$ and $(g/g_H)_{crit} = 2549.24$ (see Fig. 12.4 and 12.5). Since the presence of a vortex causes the DM density profile to drop inside the vortex core, diminishing it altogether at the very center, the effect of vortices must be seriously considered whenever the TYPE II regime is studied as a model to describe galactic halo dynamics. At the critical values for vortex formation and a little above, the effect of the vortex could be observable, as the ξ_{max}-values in Table 12.2 and 12.3 suggest. However, while it is true that the vortex becomes increasingly favored for large g (at fixed λ), its size relative to the halo size decreases, i.e. for large enough g the influence of the vortex will yet again diminish.

12.3.3 Influence on Baryonic Substructures

The foregoing sections exemplified some distinctive characteristics of Bose–Einstein-condensed DM models as compared to standard collisionless CDM. A further interesting topic is the detailed dynamics of baryonic substructure within galactic halos made of BEC-CDM. While this may likewise provide evidence for the presence or absence of this form of DM, it has hardly been considered in the previous literature.

Goodman [13] seems to be the first to discuss the consequences of TYPE II, as a superfluid, for a rotating galactic bar. This discussion has been continued in Appendix B of Slepian & Goodman [34], with the result that the drag on the bar can be smaller than estimated for standard CDM, providing a potential remedy to a problem which plagues the latter.

Lora et al. [22], on the other hand, have recently studied the dynamic survival of cold gas clumps and globular clusters in the dSph galaxies of Ursa Minor and Fornax, respectively, in both TYPE I and TYPE II regimes. The survival of those structures within the lifetime of these galaxies of known mass provides valuable constraints on the allowed boson mass and self-interaction coupling strength. For TYPE I, their calculations favor a boson mass between $0.3 \cdot 10^{-22}$ eV $<$ m $<$ 10^{-22} eV, while m can be larger for TYPE II at fixed coupling strength, their constraint reading $g/m^2 \gtrsim 8 \cdot 10^{-19}$ cm^3/eV (expressed in our units), corresponding to core sizes larger than about $R \gtrsim 0.7$ kpc. The first result can be understood given the limits in Table 12.1, while the second result is in accordance with previous studies, including our own ones as presented in Sect. 12.2. In both cases, Lora et al. [22] find viable parts of the parameter space of BEC-CDM which can explain the observations, while others can be definitely excluded.

The study of the dynamics of baryonic 'test bodies' like globular clusters or giant molecular clouds, which are large enough to feel the influence of the subtleties in the DM distribution, while still being much smaller than the DM halo, will provide useful constraints on BEC-CDM for our own Galaxy. There is valuable and interesting work to be done in future studies of this problem.

Finally, we note that galaxies are often observed to harbor supermassive black holes at their centers, with associated quasar luminosity which supports the idea that the black hole mass grew primarily by baryonic accretion. One might ask if BEC-CDM is consistent with this phenomenon, since its fluid behavior might imply a much higher accretion rate than that of collisionless CDM particles (c.f. Shapiro & Teukolsky [30]). The gravitational collapse of non-self-interacting scalar field dark matter onto a central Schwarzschild black hole in a spherically-symmetric space–time has been considered, for example, by Barranco et al. [6] and references therein, to determine if it is possible for DM halos in this model to survive for cosmological time scales. As we discussed in Sect. 12.3.2, however, halos have angular momentum. For accretion onto a central black hole to occur, this angular momentum must be overcome, which usually depends upon a dissipative process involving some form of viscosity to transfer that angular momentum outward. As a frictionless superfluid, BEC-CDM halos, however, cannot do this. Moreover, if a vortex is formed at the center, the lower density there will further inhibit accretion.

12.4 Bounds on Particle Mass and Coupling Strength

We have seen that the equilibrium size of isolated, hydrostatic BEC-CDM halo structures is predominantly governed either by quantum pressure, Eq. (12.8), or by the polytropic pressure due to the collective, repulsive self-interaction of the DM particles, Eq. (12.9), for $g \lesssim g_H$ or $g \gtrsim g_H$, respectively. By opposing gravity, quantum and self-interaction pressures each prevent structure from forming on small scales, thereby imposing lower limits on the size of structures we can expect to find in a BEC-CDM universe. The lower limit set by quantum pressure is λ_{deB} in (12.10), evaluated using the halo virial velocity (which also characterizes the bulk mass motions at its formation time). Since the smallest halos also have the smallest virial velocities, the most stringent limit results if we require $\lambda_{deB} \lesssim R$ for the smallest halos. This, in turn, imposes the lower limit on m, see Table 12.1. For a given m that satisfies this lower limit, the coupling strength g must not exceed the value such that the radius of the polytrope supported by self-interaction pressure, R_0 in (12.25) or the respective generalizations for rotating halos in (12.27) and (12.30), exceeds the size of the smallest-scale structure.

The same requirement that the characteristic size R_0 not exceed the size R of the smallest halos can also be used to place an *upper* limit on particle mass m, if we can establish an upper bound on the two-body scattering cross section per unit particle mass, σ/m, from some other argument. Slepian & Goodman [34] have argued that upper bounds on σ/m for the elastic-scattering particles in the SIDM model, based upon comparing that model to astronomical observation, should apply to BEC-CDM, as well. The interpretation of the Bullet cluster observations, for example, as a nearly collisionless merger of two cluster-sized halos has been found to limit σ/m for SIDM halos to $(\sigma/m)_{max} < 1.25$ cm^2/g, according to Randall et al. [27]. We can relate σ/m for TYPE II BEC-CDM to the characteristic size R_0 in Eq. (12.25) according to

$$\frac{\sigma}{m} = \frac{8G^2}{\pi^3 \hbar^4} R_0^4\, m^5, \tag{12.36}$$

as pointed out by Slepian & Goodman [34]. In order for (12.36) to be applicable, the particle mass must be such that $m/m_H \gg 1$, in which case RDS [28] noted that (12.36) can be rewritten in fiducial units related to the size R and mass M of a given halo or halo core, as follows:

$$\frac{\sigma}{m} = 2.094 \cdot 10^{-95} \left(\frac{m}{m_H}\right)^5 \times \left(\frac{R}{1\ \mathrm{kpc}}\right)^{3/2} \left(\frac{M}{10^8\ M_\odot}\right)^{-5/2} \frac{\mathrm{cm}^2}{\mathrm{g}}. \tag{12.37}$$

(where 'g' here means 'grams', not coupling strength). This shows that σ/m is very much smaller than $(\sigma/m)_{max} \sim 1$ cm^2/g, unless the particle mass is many orders of magnitude larger than the lower bounds, $m/m_H = 1$, in Table 12.1. We can place an *upper* bound on m, in fact, if we replace R_0 in Eq. (12.36) by R, the smallest halo (or halo core) size which must be produced, and replace σ/m by $(\sigma/m)_{max}$, to write

$$m < \left(\frac{\pi^3 \hbar^4}{8G^2}\right)^{1/5} \left(\frac{\sigma}{m}\right)_{max}^{1/5} R^{-4/5}. \tag{12.38}$$

If we take as our fiducial units $R = 1$ kpc, and $(\sigma/m)_{max} = 1\text{cm}^2/\text{g}$, this gives

$$m < 9.193 \cdot 10^{-4}\ \text{eV}/\text{c}^2. \tag{12.39}$$

To be self-consistent, we must check if our assumption is valid that BEC-CDM remains a pure condensate without thermalizing during virialization. If the relaxation time for particle collisions to establish thermodynamic equilibrium at the halo virial temperature is less than a Hubble time ($\sim 10^{17}$ sec), our assumption would break down. In that case, Slepian and Goodman [34] determined that BEC-CDM halos in the TYPE II regime would have cores surrounded by isothermal envelopes of non-condensate, which would yield density profiles in disagreement with observed halos (unless σ/m exceeds $(\sigma/m)_{max} \sim 1\ \text{cm}^2/\text{g}$ by orders of magnitude). It can be shown, however, that if $(\sigma/m)_{max} \sim 1\ \text{cm}^2/\text{g}$, then the relaxation time for achieving thermodynamic equilibrium is, indeed, more than a Hubble time, so thermodynamic equilibrium is *not* achieved for particle masses which obey inequality (12.39). Hence, this upper limit is a self-consistent one. Apparently, there is quite a large range of particle mass allowed between these upper and lower limits.

12.5 Beyond the Polytropic Size Limit

Nevertheless, this requirement that $R_0 < R$ for the smallest halos suggests there is a problem for this model in its simplest form, if observations require us to accommodate the formation of objects as small as the smallest dwarf spheroidal galaxies, while at the same time serving to explain the flattening of the density profiles in the cores of much larger galaxies. The remedy suggested by Slepian & Goodman [34], where BEC-CDM cores in the TYPE II regime are enshrouded by isothermal envelopes of non-condensate, was unfortunately shown not to work by the same authors.

We envisage a different scenario to overcome the size limit given by the equilibrium model as follows: we assume that (12.25) characterizes the size of the inner core of larger halos, which grow larger as a result of continuous infall at the time of halo formation. Let us sketch this here in more detail. It can be shown that a classic, spherical top-hat model for the collapse and virialization of a cosmological density perturbation applied to BEC-CDM in the Einstein–de Sitter universe yields a post-collapse virialized object with radius R_{TH} and uniform density ρ_0, given by

$$\frac{R_{TH}}{r_{ta}} = \frac{2}{3} \ and \ \frac{\rho_0}{\rho_{ta}} = \left(\frac{3}{2}\right)^3, \tag{12.40}$$

where r_{ta} and ρ_{ta} are the radius and density of the top-hat at the time of turn-around (i.e. maximum expansion). Then, the (virial) radius is determined from (12.22) as

$$R_{TH} = \sqrt{\frac{15}{2}} \left(\frac{g}{4\pi G m^2}\right)^{1/2}. \tag{12.41}$$

Not surprisingly, this R_{TH} depends upon g and m in the same way the polytrope radius R_0 does, so the implication is that halos of different mass must form from cosmological fluctuations that collapse at different times, in order that ρ_{ta} is different. Unfortunately, more massive halos require higher ρ_{ta}, since R_{TH} is independent of halo mass, but higher ρ_{ta} requires collapse at earlier times, which reverses the hierarchical structure formation history expected for CDM.

The crucial idea in overcoming this undesirable feature is the realization that Eq. (12.41) neglects any internal kinetic energy of the virialized object. In fact, it can be shown that the virial radius grows beyond R_{TH}, once an (effective) kinetic term has been added, even though the regime of TYPE II is retained. Determining the form and the physical meaning of this additional kinetic part, and how it can advance the above description to provide a successful model for halo formation and structure will be the subject of a forthcoming paper. An immediate observation is the fact that the effective kinetic energy K_{eff} should not simply be a constant factor times W or U_{SI}, since this will only increase the virial radius by a fixed factor times R_{TH}, as can be easily shown. In fact, we have seen that the inclusion of uniform rotation has had exactly this effect, see Eqs. (12.27) and (12.30). Instead, we will pursue the following idea: the BEC-CDM fluid must undergo oscillations during the process of virialization due to its inherently quantum-mechanical nature. In fact, the results of work by Khlopov et al. [18] and Gúzman & Ureña-López [14] lend support to this idea.

Suppose K_{eff} describes wave motions. For the general argument outlined here, it is sufficient to consider a non-vanishing bulk velocity $\mathbf{v}$ as before, i.e. $K_{eff} = T = \int \frac{\rho_0}{2}\mathbf{v}^2 dV > 0$. Since $\rho_0 = const.$, we assume that the gross average in the form of $T = \frac{3}{2}M\sigma_v^2$ with velocity dispersion $\sigma_v^2 = \frac{1}{3}\langle \mathbf{v}^2\rangle$, will capture the overall kinetic contribution due to wave motions. Additionally, in order for the virial radius to depend on halo mass and time of collapse t_{coll}, as they do for standard CDM, we require the top-hat density to be a fixed fraction of the background density at t_{coll}, that is, $\rho_0 = A\rho_{b,coll}$, with the constant A depending on the background cosmology, e.g. $A \simeq 178$ for standard CDM in a flat, matter-dominated universe. The corresponding total mass is then $M = \frac{4}{3}\pi R^3 A\rho_{b,coll}$, which yields a radius

$$R(M, t_{coll}) = \left(\frac{3M}{4\pi A}\frac{1}{\rho_{b,coll}}\right)^{1/3}. \tag{12.42}$$

Inserting this expression for R into (12.22) with $K_Q = 0$ and solving for the unknown velocity dispersion, we get

$$\sigma_v^2(M, t_{coll}) = \frac{(36\pi)^{1/3}}{15} G M^{2/3} (A\rho_{b,coll})^{1/3} - \frac{g}{2\,m^2} A\rho_{b,coll}. \tag{12.43}$$

To interpret this result, we observe the following: For $g = 0$, the last term vanishes and we recover the standard case, but for increasing $g > 0$, this term makes σ_v smaller. We can calculate the minimum mass for which $\sigma_v = 0$, resulting in

$$M_{min} = \left(\frac{15}{(36\pi)^{1/3} G}\right)^{3/2} A \left(\frac{g}{2\,m^2}\right)^{3/2} \rho_{b,coll} = \frac{4}{3}\pi A\rho_{b,coll} R_{TH,0}^3 \tag{12.44}$$

with $R_{TH,0}$ given by (12.41). Thus, the smallest mass halo has the minimum size of $R_{TH,0}$ by construction. The particle parameters may now be chosen such that M_{min} corresponds to the smallest observed galaxies, as well as to the cores of large galaxies, e.g. $M_{min} \simeq 10^8 M_{\odot}$. Larger halos then follow the relationship (12.42) and have a non-vanishing velocity dispersion due to internal wave motion, according to (12.43). This guarantees that BEC-CDM halos of mass $M > M_{min}$ would share the mass-radius relation of halos in the standard CDM model, if halos of a given mass M typically collapse at the same time as they do for standard CDM.

Acknowledgments TRD would like to thank Tonatiuh Matos and Claudia Moreno and the organizing committee for their kind hospitality at the *IV International Meeting on Gravitation and Cosmology*, Guadalajara, Mexico, May 21-25, 2012. This work was supported in part by U.S. NSF grants AST-0708176, AST-1009799 and NASA grants NNX07AH09G, NNG04G177G, NNX11AE09G to PRS. TRD also acknowledges support by the Texas Cosmology Center of the University of Texas at Austin.

References

1. K. Ahn , P.R. Shapiro, MNRAS. **363**, 1092 (2005)
2. M. Alcubierre, F.S. Guzmán, T. Matos, D. Núñez, L.A. Ureña-López, P. Wiederhold, Class. Quant. Grav. **19**, 5017 (2002)
3. A. Arbey, J. Lesgourgues, P. Salati, Phys. Rev. D **68**, 023511 (2003)
4. A. Arvanitaki, S. Dimopoulos, S. Dubovsky, N. Kaloper, J. March-Russell, Phys. Rev. D **81**, 123530 (2010)
5. J. Barnes, G. Efstathiou, ApJ. **319**, 575 (1987)
6. J. Barranco, A. Bernal, J.C. Degollado, A. Diez-Tejedor, M. Megevand, M. Alcubierre, D. Núñez, O. Sarbach, Phys. Rev. D **84**, 083008 (2011)
7. C.G. Böhmer, T. Harko, JCAP. **06**, 025 (2007)
8. S.M. Carroll, Phys. Rev. Lett. **81**, 3067 (1998)
9. S. Chandrasekhar, *Ellipsoidal Figures of Equilibrium*. (Yale Univ. Press, New Haven, 1969)
10. P.H. Chavanis, Phys. Rev. D **84**, 043531 (2011)
11. P.H. Chavanis, L. Delfini, Phys. Rev. D **84**, 043532 (2011)
12. M. Colpi, S.L. Shapiro, I. Wasserman, Phys. Rev. Lett. **57**, 2485 (1986)
13. J. Goodman, New Astronomy **5**(2), 103 (2000)
14. F.S. Guzmán, L.A. Ureña-López, 2004, Phys. Rev. D **69**, 124033 (2004)
15. U. Günther, A. Zhuk, Phys. Rev. D **56**, 6391 (1997)
16. W. Hu, R. Barkana, A. Gruzinov, Phys. Rev. Lett. **85**, 1158 (2000)
17. B. Kain, Y. Ling, Phys. Rev. D **82**, 064042 (2010)
18. M.Y. Khlopov, B.A. Malomed, Y.B. Zeldovich, MNRAS. **215** (1985)
19. J. Koda, P.R. Shapiro, MNRAS. **415**, 1125 (2011)
20. D. Lai, F.A. Rasio, S.L. Shapiro, ApJ Suppl. **88**, 205 (1993)
21. J.W. Lee, S. Lim, JCAP. **7**, 01 (2010)
22. V. Lora, J. Magaña, A. Bernal, F.J. Sánchez-Salcedo, E.K. Grebel, JCAP. **011**, 02 (2012)
23. E. Madelung, Z. für Phys. **40**, 322 (1927)
24. T. Matos, L.A. Ureña-López, Phys. Rev. D **63**, 063506 (2001)
25. M. Membrado, A.F. Pacheco, J. Sañudo, Phys. Rev. A **39**, 4207 (1989)
26. P.J.E. Peebles, ApJ. **534**(2), L127 (2000)
27. S.W. Randall, M. Markevitch, D. Clowe, A.H. Gonzalez, M. Bradač, ApJ. **679**, 1173 (2008)
28. T. Rindler-Daller, P.R. Shapiro, MNRAS. **422**, 135 (2012)

29. A. Schneider, R.E. Smith, A.V. Macciò, B. Moore, MNRAS. **424**, 684 (2012)
30. S.L. Shapiro, S.A. Teukolsky, *Black Holes, White Dwarfs, and Neutron Stars*. (Wiley, New York, 1983)
31. P. Sikivie, Q. Yang, Phys. Rev. Lett. **103**, 111301 (2009)
32. M.P. Silverman, R.L. Mallet, Gen. Rel. Grav. **34**, 633 (2002)
33. S.J. Sin, Phys. Rev. D **50**, 3650 (1994)
34. Z. Slepian, J. Goodman, MNRAS. **427**, 839 (2012)
35. A. Suárez, T. Matos, MNRAS. **416**, 87 (2011)
36. I.I. Tkachev, Phys. Lett. B **261**, 289 (1991)
37. L.A. Ureña-López, F.S. Guzmán, Phys. Rev. D **68**, 024023 (2003)
38. X.Z. Wang, Phys. Rev. D **64**, 124009 (2001)
39. T.-P. Woo, T. Chiueh, ApJ. **697**, 850 (2009)
40. R.P. Yu, M.J. Morgan, Class. Quant. Grav. **19**, L157 (2002)

Chapter 13
Inhomogeneous and Interacting Vacuum Energy

Josue De-Santiago, David Wands and Yuting Wang

Abstract Vacuum energy is a simple model for dark energy driving an accelerated expansion of the universe. If the vacuum energy is inhomogeneous in spacetime then it must be interacting. We present the general equations for a spacetime-dependent vacuum energy in cosmology, including inhomogeneous perturbations. We show how any dark energy cosmology can be described by an interacting vacuum+matter. Different models for the interaction can lead to different behaviour (e.g., sound speed for dark energy perturbations) and hence could be distinguished by cosmological observations. As an example we present the cosmic microwave microwave background anisotropies and the matter power spectrum for two different versions of a generalised Chaplygin gas cosmology.

13.1 Introduction

Vacuum energy provides a very simple model of dark energy [1–12]. It is the energy density that remains in the absence of any particles, and therefore it remains undiluted by the cosmological expansion. A positive vacuum energy can drive an accelerated expansion once the density of ordinary matter or radiation becomes subdominant. Unlike other models of dark energy it does not necessarily introduce any new dynamical degrees of freedom.

We define a vacuum energy, V, to have an energy-momentum tensor proportional to the metric

$$\check{T}^{\mu}_{\nu} = -V g^{\mu}_{\nu} \,. \tag{13.1}$$

J. De-Santiago (✉) · D. Wands · Y. Wang
Institute of Cosmology & Gravitation, University of Portsmouth,
Portsmouth, Reino Unido
e-mail: josue@ciencias.unam.mx

D. Wands
e-mail: david.wands@port.ac.uk

Y. Wang
e-mail: yuting@cs.umn.edu

C. Moreno González et al. (eds.), *Accelerated Cosmic Expansion*, Astrophysics and Space Science Proceedings 38,
DOI 10.1007/978-3-319-02063-1_13,

By comparison with the energy-momentum tensor of a perfect fluid

$$T^{\mu}_{\nu} = Pg^{\mu}_{\nu} + (\rho + P)u^{\mu}u_{\nu}\,, \tag{13.2}$$

we identify the vacuum energy density and pressure with $\check{\rho} = -\check{P} = V$, but since there is no particle flow then the four-velocity of the vacuum, $\check{u}^{\mu}$, is undefined.

A vacuum energy that is homogeneous throughout spacetime, $\nabla_{\mu} V = 0$, is equivalent to a cosmological constant in Einstein gravity, $\Lambda = 8\pi G_N V$. The discrepancy between the value of the energy density required by current observations and the typical energy scales predicted by particle physics is the long-standing cosmological constant problem [13].

We will consider the possibility of a time and/or space dependent vacuum energy. From Eq. (13.1) we have

$$\nabla_{\mu}\check{T}^{\mu}_{\nu} = Q_{\nu}. \tag{13.3}$$

where the energy flow is given by

$$Q_{\nu} \equiv -\nabla_{\nu} V. \tag{13.4}$$

We can therefore identify an *inhomogeneous vacuum*, $\nabla_{\mu} V \neq 0$, with an *interacting vacuum*, $Q_{\nu} \neq 0$ [1]. The conservation of the total energy-momentum (including matter fields and the vacuum energy) in general relativity

$$\nabla_{\mu}\left(T^{\mu}_{\nu} + \check{T}^{\mu}_{\nu}\right) = 0, \tag{13.5}$$

implies that the vacuum transfers energy-momentum to or from the matter fields

$$\nabla_{\mu}T^{\mu}_{\nu} = -Q_{\nu}. \tag{13.6}$$

Note that the energy density and four-velocity of a fluid can be identified with the eigenvalue and eigenvector of the energy-momentum tensor (13.2):

$$T^{\mu}_{\nu}u^{\nu} = -\rho u^{\mu}. \tag{13.7}$$

Because the vacuum energy-momentum tensor (13.1) is proportional to the metric tensor, any four-velocity, u^{μ}, is an eigenvector

$$\check{T}^{\mu}_{\nu}u^{\nu} = -Vu^{\mu} \quad \forall\, u^{\mu}, \tag{13.8}$$

and all observers see the same vacuum energy density, V, i.e., the vacuum energy is boost invariant.

Although the vacuum does not have a unique four-velocity, we can use the energy flow, Q_{ν}, to define a preferred unit four-vector in an inhomogeneous vacuum [1]

$$\check{u}^{\mu} = \frac{-\nabla^{\mu} V}{|\nabla_{\nu} V \nabla^{\nu} V|^{1/2}}\,. \tag{13.9}$$

normalised such that $\check{u}_\mu \check{u}^\mu = \pm 1$ for a spacelike or timelike flow. Note however that the $\check{u}^\mu$ defines a potential flow, i.e, with vanishing vorticity.

In this paper we will consider vacuum energy which may be inhomogeneous in spacetime, interacting with fields or fluids without necessarily invoking additional degrees of freedom. We show that any spatially homogeneous dark energy cosmology can be decomposed into an interacting vacuum+matter cosmology. By lifting this spatially-homogeneous solution to a covariant interaction one can study inhomogeneous perturbations which obey coupled first-order equations of motions for the matter density and velocity. As an example we consider perturbations of a generalised Chaplygin gas cosmology, decomposed into an interacting vacuum+matter.

13.2 Vacuum Cosmology

13.2.1 FRW Background

The symmetries of a spatially homogeneous and isotropic Friedmann-Robertson-Walker (FRW) metric, with scale factor $a(t)$ and Hubble rate $H = \dot{a}/a$, require the vacuum to be spatially homogeneous and isotropic too, hence $V = V(t)$. In this case the vacuum and matter are both homogeneous on spatial hypersurfaces orthogonal to the matter four-velocity, $u^\mu = (1, 0, 0, 0)$. Note that the energy flow, $\check{u}^\mu$, and matter velocity, u^μ, necessarily coincide in FRW cosmology due to the assumption of isotropy.

The Friedmann constraint equation requires

$$H^2 = \frac{8\pi G_N}{3}(\rho + V) - \frac{K}{a^2}, \tag{13.10}$$

where K determines the spatial curvature. The continuity equations for matter fields and vacuum are

$$\dot{\rho} + 3H(\rho + P) = -Q, \tag{13.11}$$

$$\dot{V} = Q. \tag{13.12}$$

The vacuum energy, V, is undiluted by the cosmological expansion, but can have a time-dependent density in the presence of a non-zero energy transfer, $Q \neq 0$, where

$$Q \equiv -u^\nu Q_\nu. \tag{13.13}$$

There have been many attempts to describe the present acceleration as due to a time-dependent vacuum energy [3–12]. However, there is little to be learnt from simply assuming an arbitrary time-dependent vacuum to obtain the desired cosmological solution. Ideally one should have a physical model from which one can derive a time-dependent solution and study other physical effects. For example, vacuum fluctuations of free fields can support an averaged density proportional to the fourth-power of the Hubble expansion, $V \propto H^4$ [14]. Such a vacuum energy would not in itself support an accelerated expansion, but other forms such as $V \propto H$ have been proposed [15] better able to match the observational data [16–20].

It is not obvious how such time-dependent vacuum models can be compared with observations in an inhomogeneous universe. We will argue that it is possible to give a consistent description of vacuum dynamics, and in particular the relativistic equations of motion for inhomogeneous perturbations, given a covariant, physical prescription for the local vacuum energy or, equivalently, the vacuum energy transfer 4-vector, $Q_\mu = -\nabla_\mu V$. One should then be able to subject vacuum models to observational constraints, even in the absence of a Lagrangian derivation or microphysical description.

13.2.2 Linear Perturbations

Let us consider inhomogeneous linear, scalar perturbations where the energy and pressure of matter is given by $\rho(t) + \delta\rho(t, x^i)$ and $P(t) + \delta P(t, x^i)$, and the four-velocity of matter is given by

$$u^\mu = \left[1 - \phi, a^{-1}\partial^i v\right], \quad u_\mu = [-1 - \phi, \partial_i \theta]. \tag{13.14}$$

where we define $\partial^i v = a(\partial x^i/\partial t)$ and $\theta = a(v + B)$. Once we allow for deviations from homogeneity in the matter and metric, we should also allow for inhomogeneity in an interacting vacuum, $V(t) + \delta V(t, x^i)$. As remarked earlier, the vacuum has an energy density and pressure, but no unique velocity. In particular the momentum of the vacuum vanishes in any frame, $(\check{\rho} + \check{P})\theta = 0$. On the other hand the energy flow $\check{u}$ defined in Eq. (13.9), can be written in analogy with the fluid velocity (13.14) as

$$\check{u}^\mu = \left[1 - \phi, a^{-1}\partial^i \check{v}\right], \quad \check{u}_\mu = \left[-1 - \phi, \partial_i \check{\theta}\right]. \tag{13.15}$$

where from Eq. (13.9) we identify $\check{\theta} = -\delta V/\dot{V}$.

Perturbations about a spatially flat ($K = 0$) FRW metric [21–24] are described by the line element

$$ds^2 = -(1 + 2\phi)dt^2 + 2a\partial_i B dt dx^i + a^2\left[(1 - 2\psi)\delta_{ij} + 2\partial_i\partial_j E\right] dx^i dx^j. \tag{13.16}$$

Following [21, 22, 24], we decompose the energy-flow along and orthogonal to the fluid velocity,

$$Q_\mu = Q u_\mu + f_\mu, \tag{13.17}$$

where $f_\mu u^\mu = 0$, so that we have

$$Q_\mu = [-Q(1 + \phi) - \delta Q, \partial_i(f + Q\theta)]\,. \tag{13.18}$$

The energy continuity equations for matter and vacuum become

$$\dot{\delta\rho} + 3H(\delta\rho + \delta P) - 3(\rho + P)\dot{\psi} + (\rho + P)\frac{\nabla^2}{a^2}\left(\theta + a^2\dot{E} - aB\right)$$
$$= -\delta Q - Q\phi, \delta\dot{V} = \delta Q + Q\phi. \tag{13.19}$$

while the momentum conservation becomes

$$(\rho + P)\dot{\theta} - 3c_s^2 H(\rho + P)\theta + (\rho + P)\phi + \delta P = -f + c_s^2 Q\theta,$$
$$-\delta V = f + Q\theta. \tag{13.20}$$

where the adiabatic sound speed $c_s^2 \equiv \dot{P}/\dot{\rho}$. Note that the vacuum momentum conservation equation becomes a constraint equation which requires that the vacuum pressure gradient is balanced by the force

$$\nabla_i(-V) = \nabla_i(f + Q\theta). \tag{13.21}$$

This determines the equal and opposite force exerted by the vacuum on the matter:

$$-f = \delta V + \dot{V}\theta. \tag{13.22}$$

i.e., the fluid element feels the gradient of the vacuum potential energy.

Note that the perturbations of a fluid coupled to the vacuum with $\check{P} = -\check{\rho}$ has no additional degrees of freedom, in contrast to a dark energy fluid with $P_X \neq -\rho_X$. Using the vacuum energy and momentum conservation equations to eliminate δQ and f we obtain

$$\dot{\delta\rho} + 3H(\delta\rho + \delta P) - 3(\rho + P)\dot{\psi} + (\rho + P)\frac{\nabla^2}{a^2}\left(\theta + a^2\dot{E} - aB\right) = -\delta\dot{V}, \tag{13.23}$$

$$(\rho + P)\dot{\theta} - 3c_s^2 H(\rho + P)\theta + (\rho + P)\phi + \delta P = \delta V + (1 + c_s^2)\dot{V}\theta. \tag{13.24}$$

13.2.3 Gauge Invariant Perturbations

It is well known that metric and matter perturbations can be gauge-dependent under a first-order gauge transformation, such as $t \to t + \delta t(t, x^i)$. The fluid density and pressure transform as $\delta\rho \to \delta\rho - \dot{\rho}\delta t$ and $\delta P \to \delta P - \dot{P}\delta t$ [24]. Similarly the vacuum perturbation transforms as $\delta V \to \delta V - Q\delta t$ and $\delta Q \to \delta Q - \dot{Q}\delta t$. The fluid 3-momentum transforms as $\theta \to \theta + \delta t$ and the energy flow transforms similarly as $\check{\theta} \to \check{\theta} + \delta t$.

We can construct gauge-invariant perturbations by specifying quantities on a particular physical reference frame [24]. For example, the vacuum perturbation on hypersurfaces orthogonal to the energy transfer, Q_μ, can be shown to vanish identically:

$$\Delta V_{\rm com} = \delta V + \dot{V}\check{\theta} = 0\,, \tag{13.25}$$

since from Eq. (13.9) and (13.15) we have $\check{\theta} = -\delta V/\dot{V}$. This simply reflects that the energy flow is the gradient of the vacuum energy and therefore the orthogonal hypersurfaces are uniform vacuum energy hypersurfaces by construction.

On the other hand the vacuum perturbation on hypersurfaces orthogonal to the matter 4-velocity, u^μ, (the *comoving vacuum perturbation*) is given by

$$\delta V_{\rm com} = \delta V + \dot{V}\theta = -f. \tag{13.26}$$

This is in general non-zero, i.e., the vacuum may be spatially inhomogeneous in the comoving-orthogonal gauge. For example, the Poisson equation for the Newtonian metric potential is given by

$$\nabla^2\Phi = 4\pi G\,(\delta\rho_{\rm com} + \delta V_{\rm com}), \tag{13.27}$$

i.e., the Newtonian metric potential is sourced by both the matter and vacuum perturbations, where

$$\delta\rho_{\rm com} = \delta\rho + \dot{\rho}\theta. \tag{13.28}$$

Note that we can write the comoving vacuum density perturbation (13.26) as

$$\delta V_{\rm com} = \dot{V}(\theta - \check{\theta}). \tag{13.29}$$

Therefore, if the energy flow follows the fluid four-velocity, $\check{u}^\mu = u^\mu$, then we have $\check{\theta} = \theta$ and the vacuum is spatially homogeneous on comoving-orthogonal hypersurfaces, $\delta V_{\rm com} = 0$.

Another gauge invariant expression for the vacuum density perturbation is the dimensionless vacuum perturbation on uniform-fluid density hypersurfaces, which describes a relative density perturbation

$$\check{S} = -3H\left(\frac{\delta V}{\dot{V}} - \frac{\delta\rho}{\dot{\rho}}\right). \tag{13.30}$$

If, for example, the vacuum energy is a function of the local matter density, $V = V(\rho)$, then the relative density perturbation must vanish and the vacuum is spatially homogeneous on uniform-density hypersurfaces, $\check{S} = 0$ (Fig. 13.1).

The total non-adiabatic pressure perturbation due to any intrinsic non-adiabatic pressure of the matter and the relative entropy perturbation between the vacuum and matter is then

$$\begin{aligned}\delta P_{\rm nad} &= \delta P - \delta V - \left(\frac{\dot{P} - \dot{V}}{\dot{\rho} + \dot{V}}\right)(\delta\rho + \delta V)\,,\\ &= \delta P - c_s^2\delta\rho + \frac{(1+c_s^2)Q[Q + 3H(\rho + P)]}{9H^2(\rho + P)}\check{S},\end{aligned} \tag{13.31}$$

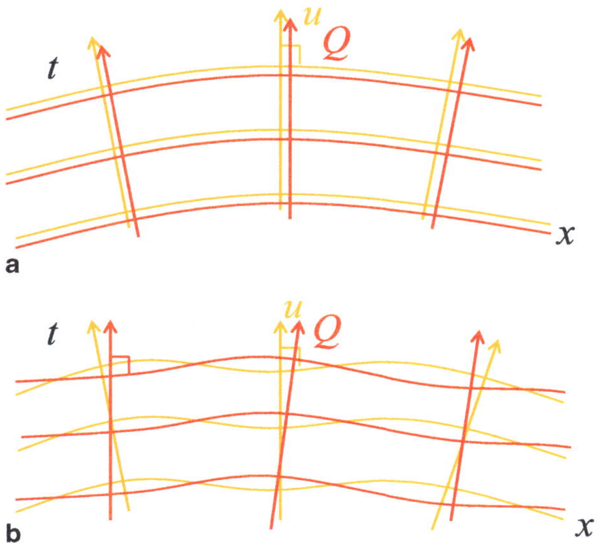

Fig. 13.1 **a** In an FRW cosmology the homogeneous spatial hypersurfaces are orthogonal to both the fluid 4-velocity, u^μ, and the vacuum energy flow, Q^μ, **b** In an inhomogeneous cosmology the spatial hypersurfaces orthogonal to the fluid 4-velocity (*light orange*) and the vacuum energy flow (*dark red*) do not necessarily coincide

where the adiabatic sound speed for matter is $c_s^2 = \dot{P}/\dot{\rho}$. This vanishes for adiabatic matter perturbations, $\delta P = c_s^2 \delta\rho$, and adiabatic vacuum fluctuations, $\check{S} = 0$, or a non-interacting vacuum, $Q = 0$.

13.3 Decomposed Generalised Chaplygin Gas

As an example of how an inhomogeneous vacuum energy might be used to describe the present accelerated expansion of our Universe we will show how one widely-studied dark energy model, the Chaplygin gas, can be re-interpreted as an interacting vacuum+matter cosmology, and how this re-interpretation can motivate different possible behaviour for density perturbations.

Any dark energy fluid energy-momentum tensor (13.2) with density $\rho_{\rm de}$ can be described by pressureless matter, with density ρ_m and velocity $u_m^\mu = u^\mu$, interacting with the vacuum, V, such that $\rho_{\rm de} = \rho_m + V$ [1]. The corresponding matter and vacuum densities are given by

$$\rho_m = \rho_{\rm de} + P_{\rm de}, \quad V = -P_{\rm de}. \tag{13.32}$$

while the energy flow is $Q_\mu = \nabla_\mu P_{\rm de}$. In an FRW cosmology this corresponds to $Q = -\dot{P}_{\rm de}$. One might choose to decompose a dark energy model $\rho_{\rm de}(a)$ into any two interacting barotropic fluids such that $\rho_{\rm de} = \rho_1 + \rho_2$, but this would double the degrees of freedom in the model unless one of these two "fluids" is the vacuum.

The generalised Chaplygin gas, defined by the barotropic equation of state [25, 26]

$$P_{\rm gCg} = -A\rho_{\rm gCg}^{-\alpha}. \tag{13.33}$$

This leads to a solution for the density in an FRW cosmology

$$\rho_{\rm gCg} = \left(A + Ba^{-3(1+\alpha)}\right)^{1/(1+\alpha)} . \tag{13.34}$$

This has the simple limiting behaviour $\rho_{\rm gCg} \propto a^{-3}$ as $a \to 0$ and $\rho_{\rm gCg} \to A^{1/(1+\alpha)}$ as $a \to +\infty$, therefore this has been proposed as a unified dark matter model. However such models are strongly constrained by observations since the barotropic equation of state defines a sound speed for matter perturbations which only reproduces the successful ΛCDM model when $\alpha \to 0$ [27, 28].

The decomposition (13.32) into pressureless matter interacting with the vacuum has previously been considered for the generalised Chaplygin gas by Bento et al. [29]. In this case we have the FRW solution

$$V = A\left(A + Ba^{-3(1+\alpha)}\right)^{-\alpha/(1+\alpha)}, \tag{13.35}$$

and hence

$$A = (\rho_m + V)^{\alpha} V. \tag{13.36}$$

The form of the FRW solution suggests a simple interaction

$$Q = 3\alpha H\left(\frac{\rho_m V}{\rho_m + V}\right). \tag{13.37}$$

In the matter or vacuum dominated limits this reduces to an interaction of the form $Q \propto H\rho_m$ or $Q \propto HV$ studied, for example, by Barrow and Clifton [30].

It is intriguing to note that the FRW solution can be defined in terms of an interaction (13.37) with a single dimensionless parameter, α, whereas when defined in terms of an equation of state (13.33) its definition requires both α and the dimensional constant A which determines the late-time cosmological constant. In the interaction model, A (and therefore the late-time cosmological constant) emerges as an integration constant dependent on initial conditions.

However, to study inhomogeneous perturbations in the decomposed model we must "lift" the explicitly time-dependent FRW solution to a covariant model for the interaction. We have at least two choices. In either case one can calculate the speed of sound in the combined interacting vacuum+matter using the usual definition [31–33, 39] and the decomposed density and pressure (13.32)

$$c_{\rm de}^2 \equiv \left(\frac{\delta P_{\rm de}}{\delta\rho_{\rm de}}\right)_{\rm com} = \left(\frac{-\delta V}{\delta\rho_m + \delta V}\right)_{\rm com} . \tag{13.38}$$

13.3.1 Barotropic Model

Firstly one could require that the local vacuum energy is a function of the local matter density, $V = V(\rho_m)$. If this applies to inhomogeneous perturbations of the interacting vacuum+matter as well as the background then we require

$$\delta V = \frac{\dot{V}}{\dot{\rho}_m}\delta\rho_m . \tag{13.39}$$

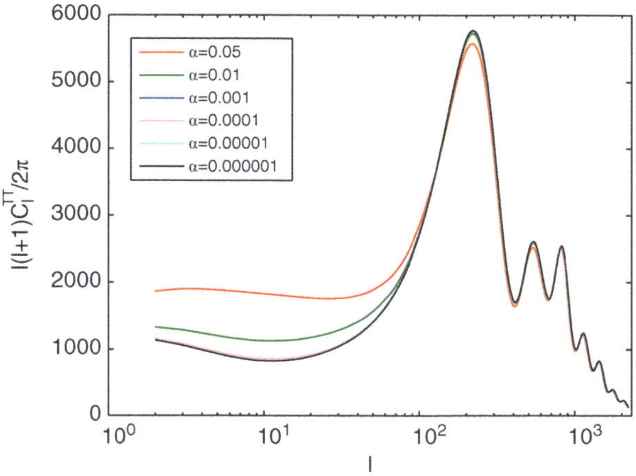

Fig. 13.2 Cosmic microwave background angular power spectrum for the generalised Chaplygin gas with barotropic equation of state and an adiabatic sound speed [34]

This implies that the vacuum perturbations are adiabatic and $\check{S} = 0$ in Eq. (13.30). On the other hand the comoving vacuum energy, (13.26), is non-zero

$$\delta V_{\text{com}} = \frac{\dot{V}}{\dot{\rho}_m} \delta\rho_{m,\text{com}}. \tag{13.40}$$

Thus one needs to consistently include the vacuum inhomogeneities as well as matter inhomogeneities, for example in the Poission Eq. (13.27), when $\dot{V} \neq 0$.

Using the adiabatic condition (13.39) we obtain the dark energy sound speed (13.38)

$$c_{\text{int}}^2 = \frac{\dot{P}_{\text{gCg}}}{\dot{\rho}_{\text{gCg}}} . \tag{13.41}$$

We see therefore the the sound speed for the interacting vacuum+fluid is the same as the adiabatic sound speed of the original barotropic Chaplygin gas (13.33). Perturbations of the the interacting vacuum+matter respect the same barotropic equation of state as the original fluid (13.33) and therefore the observational predictions for the cosmic microwave background (CMB) anisotropies, for example, are exactly the same as the original fluid model [34].

CMB temperature anisotropies as well as the power spectrum for the interacting vacuum+matter are shown in Figs. 13.2 and 13.3. We see that only very small values of the dimensionless parameter $|\alpha| < 10^{-4}$ are allowed without introducing unacceptably large oscillations in the matter power spectrum [27, 28], forcing the model to be extremely close to the standard ΛCDM cosmology.

13.3.2 Geodesic Flow

The interacting vacuum+matter model allows a different behaviour for inhomogeneous perturbations than the barotropic fluid. Suppose that the energy flow, Q_μ, is

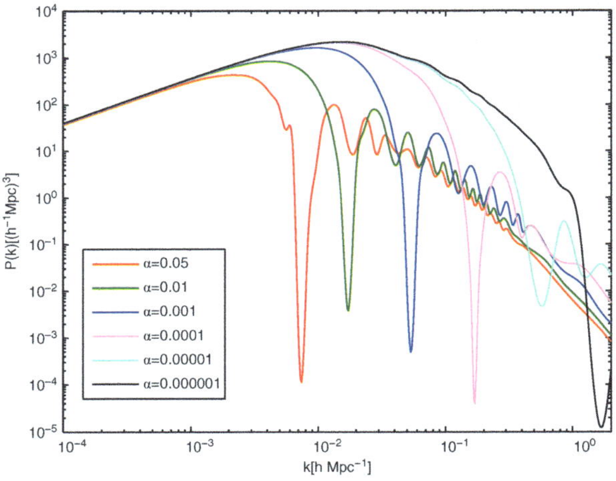

Fig. 13.3 Power spectrum for baryons plus generalised Chaplygin gas with barotropic equation of state and an adiabatic sound speed [34]

along the matter 4-velocity, u_μ. The force exert exerted by the vacuum on the matter, $-f_\mu$ in Eq. (13.17), is zero and matter 4-velocity is a geodesic flow.

In this case the comoving vacuum perturbation, $\delta V_{\rm com}$ in Eq. (13.26), is zero, i.e., the vacuum is spatially homogeneous on hypersurfaces orthogonal to the matter 4-velocity. This means that the comoving pressure perturbation for the interacting vacuum+matter is zero and hence the sound speed (13.38) is zero, $c_{\rm int}^2 = 0$ [35, 36].

The evolution of density perturbations is therefore much closer to that in a standard ΛCDM cosmology with zero sound speed than the original Chaplygin gas model with $\alpha \neq 0$. CMB temperature anisotropies as well as the matter power spectrum are shown in Figs. 13.4 and 13.5. Much larger values of $\alpha \sim 0.1$ may be compatible with the data in this case [34].

Note however that this geodesic model for the interaction does allow a gauge-invariant vacuum entropy perturbation (13.30)

$$\check{S} = \frac{3\,H}{\dot{\rho}_m}\delta\rho_{m,{\rm com}}. \tag{13.42}$$

However due to the Poisson constraint Eq. (13.27) the comoving density perturbation vanishes on large scales and hence this is compatible with adiabatic initial conditions for the interacting vacuum+matter at early times.

The requirement that $Q_\mu = Qu_\mu$ does impose an important physical restriction on the matter 4-velocity since the energy flow must then be irrotational, i.e., $u_\mu \propto \nabla_\mu V$. This would have important consquences for non-linear structure formation, possibly disastrous if gravitationally collapsed dark matter halos could not be supported by rotation. Hence we only expect geodesic matter interacting with the vacuum to be a viable model for the initial stages of structure formation and we would need a microphysical model for gravitationally collapsed halos.

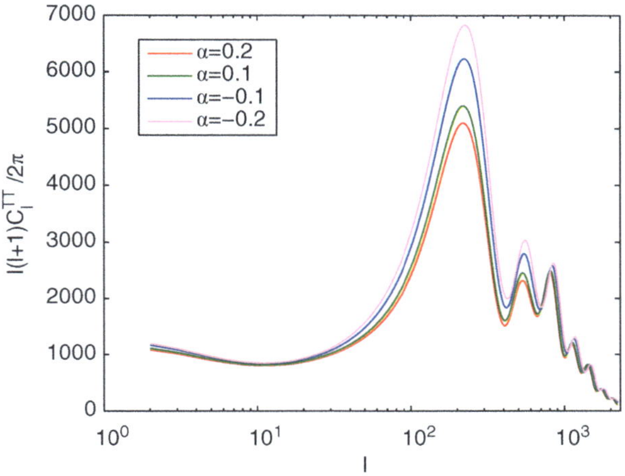

Fig. 13.4 Cosmic microwave background angular power spectrum for the decomposed generalised Chaplygin gas with geodesic matter and zero sound speed [34]

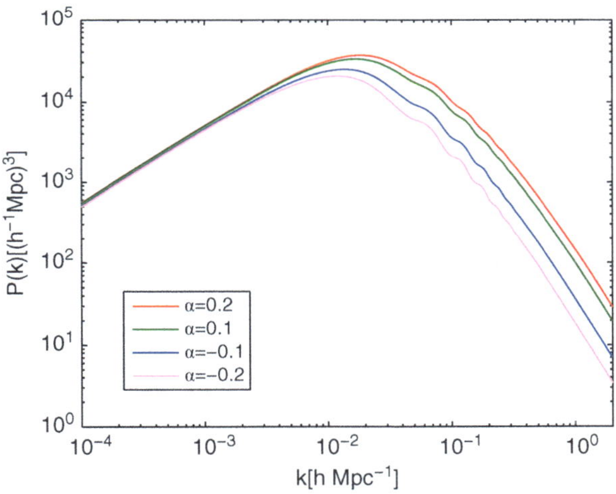

Fig. 13.5 Power spectrum for baryons plus the decomposed generalised Chaplygin gas with geodesic matter and zero sound speed [34]

13.4 Discussion

Many different explanations have been proposed for the present-day acceleration of the Universe [2]. Perhaps the most common is a self-interacting scalar field, φ, whose self-interaction potential energy could dominate the present energy density. Different models are defined by the chosen functional form for the potential energy, $V(\varphi)$ [37, 38], or the kinetic energy as a function of the field gradient, $X = (\nabla\varphi)^2$ [39, 40]. Alternative explanations include fluid models, specified by a barotropic equation of state, $P(\rho)$ [25, 41, 42], some of which could provide unified dark matter models capable of explaining acceleration on large scales and galactic dynamics (as dark matter) on much smaller scales. Acceleration requires an exotic equation of state with negative pressure, and yet the sound speed must remain real in order to avoid instabilities with respect to inhomogeneous perturbations. More sophisticated models have been proposed which allow for interactions between the different fields or fluid components, e.g., coupled quintessence models [43–48]. Such models are motivated by astronomical observations [2], providing increasingly detailed constraints on the model parameters, but the models themselves lack a persuasive underpinning physical motivation. One might therefore simply consider the simplest possible model, or parameterisation, compatible with the data.

In this paper we have considered vacuum energy as a source of spacetime curvature in Einstein gravity. A homogeneous vacuum in Einstein gravity is equivalent to a cosmological constant, whereas an inhomogeneous (time- or space-dependent) vacuum implies an interacting vacuum. Simply specifying a particular time-dependence for the vacuum energy in an FRW cosmology is unsatisfactory if the vacuum energy is introduced solely to produce a particular time-dependence of the cosmological expansion. Different physical models for the origin of the vacuum energy, or its interaction with other matter fields, will lead to different cosmological behaviour. Even models which yield the same FRW background solutions [49, 50] may be distinguished for instance by the predictions for CMB anisotropies or the matter or galaxy power spectrum.

As an example, we have shown how the generalised Chaplygin gas cosmology can be re-derived as a solution to an interacting matter+vacuum model. The interaction can be defined by a single dimensionless constant and the late-time constant vacuum energy (which appears as a dimensional parameter in the original Chaplygin gas model) can instead be derived as a constant of integration in the matter+vacuum model. We have shown the CMB and matter power spectrum predictions for two models for interacting vacuum+matter which yield the same background solution as the generalised Chaplygin gas, but give very different observational predictions [34].

Another familiar example, which we have not discussed here, would be a quintessence model with a self-interacting scalar field, φ. The self-interaction potential of the field, $V(\varphi)$, provides a vacuum energy density interacting with the kinetic energy of the scalar field. It is well known that a canonical scalar field has a sound speed equal to unity and one might thereby hope to distinguish it from a barotropic fluid model where the sound speed is necessarily equal to the adiabatic sound speed. By considering a broader class of interacting vacuum models one can study models with a range of different possible sound speeds. Our equations enable us to consider vacuum energy interacting with other forms of matter, including pressureless matter or radiation. We can therefore consider vacuum energy models which do not introduce any degrees of freedom beyond those already present in the cosmology, e.g., unified dark energy models.

In our work we have considered only linear perturbations but non-linear evolution may provide further observational constraints on different models. Often it is assumed that the vacuum energy remains unperturbed, or negligible, during collapse, but some cases, such as the interacting vacuum+matter with geodesic 4-velocity and zero sound speed may be amenable to a study of non-linear collapse [51] and we hope to study this further in future.

Acknowledgments DW is grateful to the organisers of the IVth International Conference on Gravitation and Cosmology, Guadalajara, for their hospitality. DW is supported by STFC grant ST/H002774/1. JD-S is supported by CONACYT grant 210405. JD-S and YW thank the ICG, University of Portsmouth for their hospitality.

References

1. D. Wands, J. De-Santiago, Y. Wang, Class. Quant. Grav. **29**, 145017 (2012) [arXiv:1203.6776 [astro-ph.CO]
2. E.J. Copeland, M. Sami, S. Tsujikawa, Int. J. Mod. Phys. D **15**, 1753 (2006) [hep-th/0603057]
3. O. Bertolami, Nuovo Cim. B **93**, 36 (1986)
4. K. Freese, F.C. Adams, J.A. Frieman, E. Mottola, Nucl. Phys. B **287**, 797 (1987)
5. W. Chen, S.Y. Wu, Phys. Rev. D **41**, 695 (1990) [Erratum-ibid. D **45**, 4728 (1992)]
6. J.C. Carvalho, J.A.S. Lima, I. Waga, Phys. Rev. D **46**, 2404 (1992)
7. S.M. Berman, Phys. Rev. D **43**, 1075 (1991)
8. D. Pavon, Phys. Rev. D **43**, 375 (1991)
9. A.S. Al-Rawaf, O.M. Taha, Phys. Lett. B **366**, 69 (1996)
10. I. L. Shapiro, J. Sola, JHEP **0202**, 006 (2002) [hep-th/0012227]
11. T. Buchert, N. Obadia, Class. Quant. Grav. **28**, 162002 (2011) [arXiv:1010.4512 [gr-qc]
12. J. Sola, J. Phys. Conf. Ser. **283**, 012033 (2011) [arXiv:1102.1815 [astro-ph.CO]]
13. S. Weinberg, Rev. Mod. Phys. **61**, 1 (1989)
14. T.S. Bunch, C.W. P. Davies, Proc. Roy. Soc. Lond. A **360**, 117 (1978)
15. R. Schutzhold, Phys. Rev. Lett. **89**, 081302 (2002)
16. J. C. Fabris, I. L. Shapiro, J. Sola, JCAP **0702**, 016 (2007) [gr-qc/0609017]
17. M.A. Velasquez-Toribio, Int. J. Mod. Phys. D **21**, 1250026 (2012) [arXiv:0907.3518 [astro-ph.CO]]
18. W. Zimdahl, H.A. Borges, S. Carneiro, J.C. Fabris, S. W. Hipolito-Ricaldi, JCAP **1104**, 028 (2011) [arXiv:1009.0672 [astro-ph.CO]
19. L. Xu, Y. Wang, M. Tong, H. Noh, Phys. Rev. D **84**, 123004 (2011) [arXiv:1112.5216 [astro-ph.CO]]
20. J.S. Alcaniz, H.A. Borges, S. Carneiro, J.C. Fabris, C. Pigozzo, W. Zimdahl, [arXiv:1201.5919] [astro-ph.CO]
21. K.A. Malik, D. Wands, JCAP **0502**, 007 (2005) [astro-ph/0411703]
22. H. Kodama, M. Sasaki, Prog. Theor. Phys. Suppl **78**, 1 (1984)
23. V.F. Mukhanov, H.A. Feldman, H.R. Brandenberger, Phys. Rept **215**, 203 (1992)
24. K.A. Malik, D. Wands, Phys. Rept **475**, 1 (2009) [arXiv:0809.4944 [astro-ph]].
25. A.Y. Kamenshchik, U. Moschella, V. Pasquier, Phys. Lett. B **511**, 265 (2001) [gr-qc/0103004]
26. M.C. Bento, O. Bertolami, A.A. Sen, Phys. Rev. D **66**, 043507 (2002) [gr-qc/0202064].
27. H. Sandvik, M. Tegmark, M. Zaldarriaga, I. Waga, Phys. Rev. D **69**, 123524 (2004) [astro-ph/0212114]
28. C. -G. Park, J. -c. Hwang, J. Park, H. Noh, Phys. Rev. D **81**, 063532 (2010) [arXiv:0910.4202 [astro-ph.CO]]
29. M.C. Bento, O. Bertolami, A.A. Sen, Phys. Rev. D **70**, 083519 (2004) [astro-ph/0407239]
30. J. D. Barrow, T. Clifton, Phys. Rev. D **73**, 103520 (2006) [gr-qc/0604063]
31. J. Weller, M.A. Lewis, Mon. Not. Roy. Astron. Soc **346**, 987 (2003) [astro-ph/0307104]
32. R. Bean, O. Dore, Phys. Rev. D **69**, 083503 (2004) [astro-ph/0307100].
33. S. Hannestad, Phys. Rev. D **71**, 103519 (2005) [astro-ph/0504017]
34. Y. Wands, D. Wands, L. Xi, J. De-Santiago, A. Hojjati, Phys. Rev. D **87**, 083503 (2013) [arXiv:1301.5315 [astro-ph.CO]]
35. P. Creminelli, G. D'Amico, J. Norena, F. Vernizzi, JCAP **0902**, 018 (2009) [arXiv:0811.0827 [astro-ph]

36. E. A. Lim, I. Sawicki and A. Vikman, JCAP **1005**, 012 (2010) [arXiv:1003.5751 [astro-ph.CO]
37. B. Ratra, J.E.P. Peebles, Phys. Rev. D **37**, 3406 (1988)
38. R.R. Caldwell, R. Dave, J.P. Steinhardt, Phys. Rev. Lett. **80**, 1582 (1998) [astro-ph/9708069]
39. C. Armendariz-Picon, T. Damour, F.V. Mukhanov, Phys. Lett. B **458**, 209 (1999) [hep-th/9904075]
40. C. Armendariz-Picon, V.F. Mukhanov, J.P. Steinhardt, Phys. Rev. D **63**, 103510 (2001) [astro-ph/0006373]
41. E.V. Linder, J.R. Scherrer, Phys. Rev. D **80**, 023008 (2009) [arXiv:0811.2797 [astro-ph]]
42. A. Balbi, M. Bruni, C. Quercellini, Phys. Rev. D **76**, 103519 (2007) [astro-ph/0702423]
43. C. Wetterich, Astron. Astrophys. **301**, 321 (1995) [hep-th/9408025]
44. L. Amendola, Phys. Rev. D **62**, 043511 (2000) [astro-ph/9908023]
45. D.J. Holden, D. Wands, Phys. Rev. D **61**, 043506 (2000) [gr-qc/9908026]
46. A.P. Billyard, A.A. Coley, Phys. Rev. D **61**, 083503 (2000) [astro-ph/9908224]
47. G.R. Farrar, J.E. P. Peebles, Astrophys. J **604**, 1 (2004) [astro-ph/0307316]
48. C.G. Boehmer, G. Caldera-Cabral, R. Lazkoz, R. Maartens, Phys. Rev. D **78**, 023505 (2008) [arXiv:0801.1565 [gr-qc]]
49. M. Kunz, Phys. Rev. D **80**, 123001 (2009) [astro-ph/0702615]
50. A. Aviles, L.J. Cervantes-Cota, Phys. Rev. D **84**, 083515 (2011) [Erratum-ibid. D **84**, 089905 (2011)] [arXiv:1108.2457 [astro-ph.CO]]
51. P. Creminelli, G.D' Amico, J. Norena, L. Senatore, F. Vernizzi, JCAP **1003**, 027 (2010) [arXiv:0911.2701 [astro-ph.CO]]

Part III
Other Topics

Chapter 14
Relativistic Dust Thin Disks with Halo and Magnetic Field

Diego A. Ballén-Daza, Guillermo A. González and Antonio C. Gutiérrez-Piñeres

Abstract A new family of exact solutions of the Einstein-Maxwell equations is presented, which describes an infinitely axially symmetric thin disk surrounded by a spheroidal halo with magnetic field. The models are obtained from axisymmetric solutions of Einstein-Maxwell equations for conformastatic spacetimes in which the metric and the magnetic potential present a discontinuity in its normal first derivative through a thin disk. The energy-momentum tensor and the current density are found and expressed in terms of the magnetic potential and the metric function and the system satisfies all the energy conditions and the total mass it is finite. Finally, we consider a particular case of Kuzmin-Toomre disks, by showing the behavior of the four-vector current and energy density profile for the halo as well as for disk.

14.1 Introduction

The study of disklike structures based on axially symmetric solutions of the Einstein-Maxwell equations gain relevance from observational evidence, which proves that such configurations are diverse in objects and astrophysical phenomena. In adition, although in these structures the most of energy is in a flat disk, there is no doubt that exists a region of space called *halo*, which wraps the disk and it is composed of baryonic matter and maybe dark matter. On the other hand, the interest in considering magnetic fields is motivated by the presence in the nature of many astrophysical objects with strong magnetic field. Numerous observations, for instance, reveal the existence of magnetic fields in the disk of our Galaxy. Since then, our solutions may be used to model some kinds of galaxies in thermodynamic equilibrium or accretion disks around black holes (see, for instance [1]).

D. A. Ballén-Daza (✉) · G. A. González
Escuela de Física, Universidad Industrial de Santander,
Bucaramanga, Colombia
e-mail: diegoballen@gmail.com

G. A. González
e-mail: guillego@uis.edu.co

A. C. Gutiérrez-Piñeres
Instituto de Ciencias Nucleares, UNAM,
Mexico, Colombia
e-mail: acgutierrez@nucleares.unam.mx

C. Moreno González et al. (eds.), *Accelerated Cosmic Expansion*,
Astrophysics and Space Science Proceedings 38,
DOI 10.1007/978-3-319-02063-1_14,

14.2 Models of Thin Disks and Halos

We take the metric tensor as given by the conformastatic line element [2], written in cylindrical coordinates $x^a = (t, \varphi, r, z)$ as,

$$ds^2 = -e^{2\psi} dt^2 + e^{-2\psi} \left(r^2 d\varphi^2 + dr^2 + dz^2\right), \tag{14.1}$$

where the metric function ψ depends only on r and z. The azimutal angle φ ranges in the usual interval $[0, 2\pi)$, while the coordinate r increases in the interval $(0, \infty)$ and the coordinate z does in $(-\infty, \infty)$. Now, we assume that in the spacetime there exists a thin disk, located at the hypersurface $z = 0$, such that the components of the metric tensor $g_{\mu\nu}$ and electromagnetic potential A_μ are symmetrical functions of z, and their first derivatives have a finite discontinuity at $z = 0$. Accordingly,

$$g_{\mu\nu}(r, z) = g_{\mu\nu}(r, -z), \tag{14.2}$$

$$A_\mu(r, z) = A_\mu(r, -z), \tag{14.3}$$

in such way that, for $z \neq 0$,

$$g_{\mu\nu,z}(r, z) = -g_{\mu\nu,z}(r, -z), \tag{14.4}$$

$$A_{\mu,z}(r, z) = -A_{\mu,z}(r, -z). \tag{14.5}$$

Therefore, the metric tensor and the electromagnetic potential are continuous at $z = 0$,

$$\left[g_{\mu\nu}\right] = g_{\mu\nu}{}^{+}|_{z=0^+} - g_{\mu\nu}{}^{-}|_{z=0^-} = 0, \tag{14.6}$$

$$\left[A_\mu\right] = A_\mu{}^{+}|_{z=0^+} - A_\mu{}^{-}|_{z=0^-} = 0, \tag{14.7}$$

where the superscrpts $+$ and $-$ represent the region above and below of $z = 0$ respectively. The expression "[ ]" denotes the jump of the tensor field. The discontinuity in the derivatives of the metric tensor and the electromagnetic potential can be written as,

$$b_{\mu\nu} = \left[g_{\mu\nu,z}\right] = 2g_{\mu\nu,z}|_{z=0^+}, \tag{14.8}$$

$$t_\mu = \left[A_{\mu,z}\right] = 2A_{\mu,z}|_{z=0^+}, \tag{14.9}$$

where it has been used the antisymmetry of reflection with respect to the hypersurface in $z = 0$. Then, by using the distributional approach [3] we can write the metric and electromagnetic potential as

$$g_{\mu\nu} = (g_{\mu\nu})^D = g^{+}_{\mu\nu}\Theta(z) + g^{-}_{\mu\nu}\{1 - \Theta(z)\}, \tag{14.10}$$

$$A_\mu = (A_\mu)^D = A^{+}_\mu\Theta(z) + A^{-}_\mu\{1 - \Theta(z)\}, \tag{14.11}$$

being $\Theta(z)$ the Heaviside function. The letter D indicates a distribution, i.e, this tensorial field is distributed in the above region as well as below region of a thin disk. Hence, the Ricci tensor is written as

$$R_{\mu\nu} = R^{+}_{\mu\nu}\Theta(z) + R^{-}_{\mu\nu}\{1 - \Theta(z)\} + H_{\mu\nu}\delta(z), \tag{14.12}$$

where $\delta(z)$ is the Dirac distribuion with support on $z = 0$. Here, $R^{\pm}_{\mu\nu}$ represents the Ricci tensor in the regions $z < 0$ and $z > 0$, and $H_{\mu\nu}$ is the Ricci tensor associated to disk and it is given by

$$H_{\mu\nu} = \frac{1}{2}\left(b^{z}{}_{\mu}\delta^{z}_{\nu} + b^{z}{}_{\nu}\delta^{z}_{\mu} - b^{\alpha}{}_{\alpha}\delta^{z}_{\mu}\delta^{z}_{\nu} - g^{zz}b_{\mu\nu}\right). \tag{14.13}$$

Likewise, from Eqs. (14.9) y (14.11), we obtain

$$F^{\mu\nu}{}_{,\nu} = \left(F^{\mu\nu}{}_{,\nu}\right)^{D} + \left[F^{\mu\nu}\right] n_{\nu}\delta\,(z), \tag{14.14}$$

where n_{ν} is a unit normal vector to hypersurface and $F^{\mu\nu}$ is the Faraday tensor, which is defined in its covariant form as

$$F_{\mu\nu} = A_{\nu,\mu} - A_{\mu,\nu}. \tag{14.15}$$

In order to consider only magnetic fields, the electromagnetic potential has the form $A_{\mu} = (0, A, 0, 0)$, where A is the vector potential.

Now, we must to solve the Einstein-Maxwell equations, which in geometrized units such that $c = G = \mu_0 = \varepsilon_0 = 1$, can be expressed as

$$G_{\mu\nu} = 8\pi T_{\mu\nu}, \tag{14.16a}$$

$$F^{\mu\nu}{}_{;\nu} = 4\pi J^{\mu}, \tag{14.16b}$$

where J^{μ} is the current four-vector, $T_{\mu\nu}$ is the energy-momentum tensor and $G_{\mu\nu}$ is the Einstein tensor given by

$$G_{\mu\nu} = R_{\mu\nu} - \frac{1}{2}g_{\mu\nu}R, \tag{14.17}$$

where R is the Ricci scalar. From Eq. (14.12) and (14.14), the current four-vector and the energy-momentum tensor can be written as

$$T_{\mu\nu} = (T_{\mu\nu})^{D} + Q_{\mu\nu}\delta(z), \tag{14.18}$$

$$J^{\mu} = (J^{\mu})^{D} + I^{\mu}\delta(z), \tag{14.19}$$

where $Q_{\mu\nu}$ and I^{μ} are the energy-momentum tensor and current four-vector associated to the hypersurface on $z = 0$, and $\left(T_{\mu\nu}\right)^{D}$ and $(J^{\mu})^{D}$ correspond to the halo.

Then, the Einstein-Maxwell equations are equivalent to

$$8\pi T_{\mu\nu}^{\pm} = R_{\mu\nu}^{\pm} - \frac{1}{2} g_{\mu\nu} R^{\pm}, \tag{14.20a}$$

$$F_{\pm\ ;\nu}^{\mu\nu} = 4\pi J_{\pm}^{\mu}, \tag{14.20b}$$

$$8\pi Q_{\mu\nu} = H_{\mu\nu} - \frac{1}{2} g_{\mu\nu} H, \tag{14.20c}$$

$$[F^{\mu\nu}]n_{\nu} = 4\pi I^{\mu}, \tag{14.20d}$$

where (14.20a) and (14.20b) represent the halo, while (14.20c) and (14.20d) represent the disk.

We consider that the energy-momentum tensor of the halo can be expressed as

$$T_{\mu\nu}^{\pm} = \overset{m}{T_{\mu\nu}^{\pm}} + \overset{em}{T_{\mu\nu}^{\pm}}, \tag{14.21}$$

where the first term on right hand corresponds to material content (m) and the second term to electromagnetic content (em), whish is defined by

$$\overset{em}{T_{\mu\nu}^{\pm}} = \frac{1}{4\pi}\left\{F_{\mu\alpha}F^{\alpha}{}_{\nu} - \frac{1}{4}g_{\mu\nu}F_{\alpha\lambda}F^{\alpha\lambda}\right\}, \tag{14.22}$$

so that, we can find the material content, since $T_{\mu\nu}^{\pm}$ is obtained from Einstein tensor. Before that, we propose an ansatz, which relates the metric function and the magnetic potential. This relationship is given by

$$A_{,r} = \pm kre^{-\psi}\psi_{,z}, \tag{14.23a}$$

$$A_{,z} = \mp kre^{-\psi}\psi_{,r}, \tag{14.23b}$$

where k is a constant. The integrability condition of the above system it is granted by

$$\nabla^2\psi = \nabla\psi \cdot \nabla\psi. \tag{14.24}$$

Taking into account the above equations, the material content of the energy-momentum tensor is

$$8\pi \overset{m}{T_{00}^{\pm}} = e^{4\psi}\left(1-k^2\right)\nabla\psi \cdot \nabla\psi, \tag{14.25a}$$

$$8\pi \overset{m}{T_{11}^{\pm}} = r^2\left(1-k^2\right)\nabla\psi \cdot \nabla\psi, \tag{14.25b}$$

$$8\pi \overset{m}{T_{22}^{\pm}} = \left(1-k^2\right)\left(\psi_{,z}{}^2 - \psi_{,r}{}^2\right), \tag{14.25c}$$

$$8\pi \overset{m}{T_{33}^{\pm}} = \left(1-k^2\right)\left(\psi_{,r}{}^2 - \psi_{,z}{}^2\right), \tag{14.25d}$$

$$8\pi \overset{m}{T_{32}^{\pm}} = 8\pi \overset{m}{T_{23}^{\pm}} = -2\left(1-k^2\right)\left(\psi_{,z}\psi_{,r}\right), \tag{14.25e}$$

where all remaining components are zero. Also, we find that the current density is

$$\widehat{J_{\pm}}^1 = 0, \tag{14.26}$$

where $\widehat{J_{\pm}}^\mu = J_{\pm}{}^\mu \sqrt{-g}$, being g the determinant of the metric tensor.

The only nonzero component of the energy-momentum tensor of disk is given by

$$Q_{00} = \frac{e^{4\psi}\,\psi_{,z}}{2\pi}, \tag{14.27}$$

where all remaining components are zero. Accordingly, the only nonzero component of the surface energy-momentum tensor, defined as

$$S_{\mu\nu} = \int E_{\mu\nu}\, ds_n = \sqrt{g_{zz}}\, Q_{\mu\nu}, \tag{14.28}$$

where $E_{\mu\nu} = Q_{\mu\nu}\delta(z)$ and $ds_n = \sqrt{g_{zz}}\ dz$ is the "physical measure" of lenght in the normal direction to the hypersurface on $z = 0$, is given by

$$S_{00} = \frac{e^{3\psi}\,\psi_{,z}}{2\pi}, \tag{14.29}$$

so the disk describes a dust material. For the surface current density we obtain,

$$I^1 = -k\frac{e^{3\psi}}{2\pi r}\psi_{,r}|_{z=0^+}. \tag{14.30}$$

while the remaining components are zero.

14.3 Physical Content of the Model

In order to analyze the physical characteristics of the system, we will use the tetrad of the "locally tatic observer" [4], given by

$$e^\mu{}_{(0)} = e^{-\psi}\,\delta^\mu_0, \tag{14.31a}$$

$$e^\mu{}_{(1)} = \frac{e^{\psi}}{r}\delta^\mu_1, \tag{14.31b}$$

$$e^\mu{}_{(2)} = e^{\psi}\,\delta^\mu_2, \tag{14.31c}$$

$$e^\mu{}_{(3)} = e^{\psi}\,\delta^\mu_3, \tag{14.31d}$$

where $e^{\mu}{}_{(0)}$ is the four-velocity. Then, after diagonalize the energy-momentum tensor, we obtain

$$8\pi\rho = e^{2\psi}\left(1-k^2\right)\nabla\psi\cdot\nabla\psi, \tag{14.32a}$$

$$8\pi P_1 = e^{2\psi}\left(1-k^2\right)\nabla\psi\cdot\nabla\psi, \tag{14.32b}$$

$$8\pi P_2 = e^{2\psi}\left(1-k^2\right)\nabla\psi\cdot\nabla\psi, \tag{14.32c}$$

$$8\pi P_3 = -e^{2\psi}\left(1-k^2\right)\nabla\psi\cdot\nabla\psi, \tag{14.32d}$$

where ρ is the energy density of halo and P_1, P_2 and P_3 are the pressures, so that

$$P = \frac{P_1+P_2+P_3}{3} = \frac{\rho}{3}, \tag{14.33}$$

is the average pressure. Accordingly, by impossing over the constant k the restriction

$$-1 < k < 1, \tag{14.34}$$

the energy-momentum tensor will be in agreement with all the energy conditions [5].

Likewise, for the surface energy density of the disk we obtain

$$\sigma = S_{(0)(0)} = \frac{e^{\psi}\,\psi_{,z}}{2\pi}, \tag{14.35}$$

while the current density is

$$I^{(1)} = -k\frac{e^{2\psi}}{2\pi}\psi_{,r}, \tag{14.36}$$

which corresponds to the azimuthal component.

The components of the magnetic field as measured by the static observer are given by

$$B_{(r)} = -\frac{e^{2\psi}A_{,z}}{r}, \tag{14.37a}$$

$$B_{(z)} = \frac{e^{2\psi}A_{,r}}{r}, \tag{14.37b}$$

$$B_{(\varphi)} = 0, \tag{14.37c}$$

and the components of the observer acceleration are

$$a_{(r)} = e^{\psi}\,\psi_{,r}, \tag{14.38a}$$

$$a_{(z)} = e^{\psi}\,\psi_{,z}, \tag{14.38b}$$

so that, using Eq. (14.23a) and (14.23b), we obtain

$$kB_{(r)} = a_{(r)}, \tag{14.39a}$$

$$kB_{(z)} = a_{(z)}, \tag{14.39b}$$

and thus, the gravitational and the electromagnetic fields are aligned. The magnetic field lines can be obtained by solving the differential equation

$$\frac{dz}{B_{(z)}} = \frac{dr}{B_{(r)}}, \tag{14.40}$$

that using Eqs. (14.37a) and (14.37b) can be written as

$$dA = A_{,r}\, dr + A_{,z}\, dz = 0. \tag{14.41}$$

Accordingly, the equation

$$A(r, z) = C, \tag{14.42}$$

with C constant, can be used to obtain the lines of force both of the magnetic field as of the gravitational field.

The integrabilty condition (14.24) can be equivalently written as

$$\nabla^2(e^{-\psi}) = 0, \tag{14.43}$$

in such way that, in order to have an asymptotically flat spacetime, we can write

$$e^{-\psi} = 1 - \Phi, \tag{14.44}$$

where Φ is a solution of the Laplace equation,

$$\nabla^2\Phi = 0, \tag{14.45}$$

and we shall consider only those solutions that vanishes at infinite.

Substituting the above expressions in the energy and current densities of the halo and disk, we obtain

$$\rho = \frac{1-k^2}{\left(\Phi-1\right)^4}\ \nabla\Phi \cdot \nabla\Phi, \tag{14.46}$$

$$\hat{J}^1 = 0, \tag{14.47}$$

$$\sigma = \frac{1}{2\pi\left(\Phi-1\right)^2}\ \Phi_{,z}, \tag{14.48}$$

$$I^{(1)} = -\frac{k}{2\pi\left(1-\Phi\right)^3}\ \Phi_{,r}, \tag{14.49}$$

where the Eqs. (14.46) and (14.47) correspond to the halo, and Eqs. (14.48) and (14.49) to the disk.

14.4 The Kuzmin-Toomre Disks

In this section we consider a particular solution of the Laplace equation given by

$$\Phi_n(R,\theta) = -\sum_{l=0}^{n} \frac{C_l}{R^{l+1}}\ P_l(cos\theta), \tag{14.50}$$

where C_l are constants, P_l $(cos\theta)$ are the Legendre Polynomials [6] and,

$$R = \sqrt{r^2 + z^2}. \tag{14.51}$$

Now, as this solution and all its derivatives are continuous, we introduce a discontinuityby means of the transformation

$$z \longrightarrow |z| + d, \tag{14.52}$$

where d is a positive constant, leading so to the well known Kuzmin & Toomre [7] family of disks.

In this paper we are going to work out the solution in Eq. 14.50 for $n = 0$,

$$\Phi_0 = -\frac{C_0}{R}, \tag{14.53a}$$

$$\Phi_{0,r} = r\ \frac{C_0}{R^3}, \tag{14.53b}$$

$$\Phi_{0,z} = z\ \frac{C_0}{R^3}. \tag{14.53c}$$

Then, for the energy density of the halo we obtain

$$\tilde{\rho} = \frac{1}{(1+\tilde{R})^4}, \tag{14.54}$$

where $\tilde{\rho} = \rho/\rho_0$, $\tilde{R} = R/C_0$ and $\rho_0 = (1-k^2)/{C_0}^2$. In order to find the total mass of the system, we use the Komar mass [5, 8],

$$M = 2\int_{\Sigma}\left(T_{\mu\nu} - \frac{1}{2}T g_{\mu\nu}\right) n^{\mu} \xi^{\nu}{}_{(t)} \sqrt{h}\ d^3 y, \tag{14.55}$$

from what we can find a general expression for the halo mass, that for model $n = 0$ gives

$$M_{H_0} = \frac{4\pi(1-k^2){C_0}^2}{d}. \tag{14.56}$$

Now, for the surface energy density of the disk we find

$$\tilde{\sigma} = \frac{1}{\sqrt{\tilde{r}^2+1}}\left[\frac{\tilde{C}_0 + 1}{\tilde{C}_0 + \sqrt{\tilde{r}^2+1}}\right]^2, \tag{14.57}$$

where $\tilde{r} = r/d$, $\tilde{C}_0 = C_0/d$ and $\tilde{\sigma} = \sigma/\sigma_0$, with

$$\sigma_0 = \frac{C_0}{2\pi(C_0 + d)^2}, \tag{14.58}$$

and all the previous expressions are evaluated on $z = 0^+$.

The disk mass reduces to

$$M_{D_0} = \ln\left(1 + \frac{C_0}{d}\right). \tag{14.59}$$

in such a way that the total mass is given by

$$M_T = \frac{4\pi(1 - k^2){C_0}^2}{d} + \ln\left(1 + \frac{C_0}{d}\right). \tag{14.60}$$

The surface current density is

$$I^{(1)} = \lambda \frac{\tilde{C}_0 \tilde{r}}{(\sqrt{\tilde{r}^2 + 1} + \tilde{C}_0)^3}, \tag{14.61}$$

where $\lambda = -k/2\pi d$, $\tilde{C}_0 = C_0/d$, $\tilde{r} = r/d$ and all expressions are evaluated on $z = 0^+$.

For the magnetic potential we find

$$A(r, z) = \frac{k\tilde{C}_1 \tilde{r}^2}{\tilde{R}^3} - \frac{kC_0(|z| + 1)}{\tilde{R}} = C, \tag{14.62}$$

where $\tilde{r} = r/d$, $\tilde{C}_1 = C_1/d$, $\tilde{z} = z/d$ and $\tilde{R} = R/d$ and C is a constant. Then, we can write a equation for the magnetic field lines,

$$\tilde{R}^3 = \epsilon \tilde{r}^2 - \kappa(|\tilde{z}| + 1)\tilde{R}^2, \tag{14.63}$$

where $\epsilon = k\tilde{C}_1/C_B$ and $\kappa = kC_0/C_B$. In Figs. 14.1 we present the behavior and physical intrpretation of volumetric energy density in (14.54); surface energy density in (14.57); surface current density in (14.61) and the magnetic field lines given by (14.63).

14.5 Concluding Remarks

We obtained a new infinite family of exact solutions of the Einstein-Maxwell equations for a static and axially symmetric spacetime. These solutions describe a family of systems consisting of a thin disk, surrounded by a spheroidal halo of matter in presence of magnetic fields. Although the disk is infinite, the volumetric and surface energy density show a suitable behavior, i.e., its maximum value is at the center of

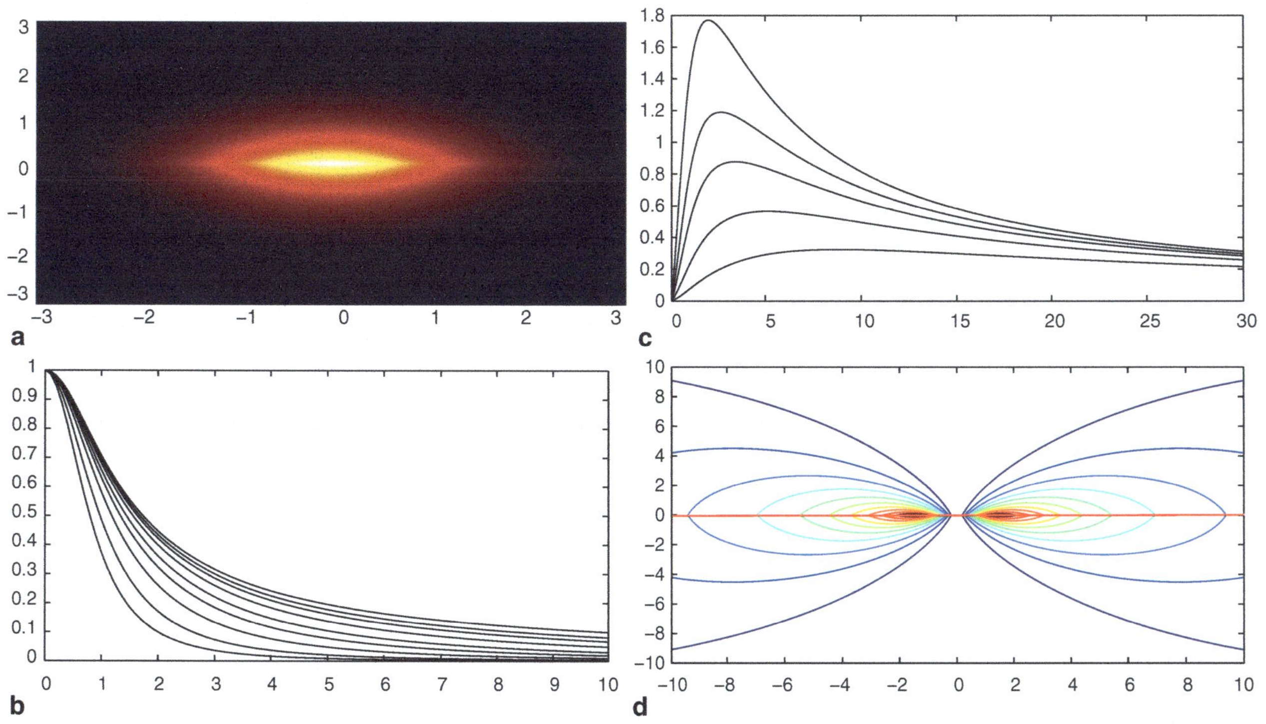

Fig. 14.1 **a)** We show, for the model $n = 0$, the volumetric energy density $\tilde{\rho}$ as function of $\tilde{r}$ and $\tilde{z}$ where the most intense colors indicate the regions where the energy density is high. We can see that energy density decreases quickly, doing evident an asymptotically flat spacetime given by (14.44). **b)** We show the surface energy density $\tilde{\sigma}$ as function of $\tilde{r}$ for different values of $\tilde{C}_0$. Beginning from the bottom to top curve $\tilde{C}_0 = 0.1; 1; 3; 6; 10; 20; 40; 80; 3000$. Likewise, we can see that energy density decreases quickly, doing evident an asymptotically flat spacetime given by (14.44). **c)** The surface current density $I^{(1)}$ is then shown as function of $\tilde{r}$, for $\lambda = 10$ and different values of C_0. Beginning from the top to the bottom curve, $\tilde{C}_0 = 1; 1.5; 2; 3; 5$. **d)** Finally, we plot $\tilde{z}$ as function of $\tilde{r}$ in order to show the magnetic field lines, with $\epsilon = 5000$ and $\kappa = 2$. The horizontal red line shows the edge disk in $\tilde{z} = 0$. In regions where this lines are closer, the surface current density is higher

the system and vanishes rapidly as the distance from the center increases. From the expressions for the energy-momentum tensor, we studied the physical behavior of the model, by verifying compliance with the energy conditions. Since the introduction of the magnetic field to the model, it was found that the charge density of the halo and the disk is zero, which agrees with the observational evidence, showing in general that galaxies and other astrophysical objects are essentially neutral. We found that the material content of disk describes *dust*, which is given by the unique component in the energy-momentum tensor of disk in Eq. (14.27). This shows that this tensor lacks of pressure, since other components in diagonal are null. It was also found that total mass of the model is finite, which is achieved by ensuring that the solution describes an asymptotically flat spacetime.

References

1. M. Rees, in *Black Holes and Relativistic Stars*, Ed by R. (Wald University o Chicago Press, Chicago, 1998), p. 79.
2. L. Synge, in *Relativity, the General Theory*, (North-Holland Publishing Company, Amsterdam, 1966)
3. A. Papapetrou y A. Hamouni, Ann. Ins. Henri. Poincaré **9**, 179–188 (1968)
4. J. Katz, J. Biucák y Lynden-Bell D, Class. Quantum Grav **21**, 1–51 (2004)
5. R. Wald, *General Relativity*. (The University of Chicago Press, Chicago y Londres, 1984)
6. G. Arfken, H. Weber, *Mathematical Methods for Physicists*. (Academic Press, New York 2005)
7. J. Binney, S. Tremaine, *Galactic Dynamics*, 2nd ed. (Princeton University Press, 2008)
8. C. Misner, K. Thorne, J. Wheeler, *Gravitation* (Freeman and Company 1970)
9. T.M. Apostol, *Calculus*, Reverté S.A. (1965)
10. G.A. González y P.S. Letelier, Class. Quantum Grav *16*, 479 (1999)
11. G.A. González, A.C. Gutierrez-Piñeresy V.M., Phys. Rev. D **79**, 124048, 2009
12. G. García-Reyes y G.A. González, Phys. Rev D. **70**, 104005 (2004)
13. G.A. González, A.C. Gutierrez-Piñeresy P, Phys. Rev. D **78**, 064058 (2008)
14. A.C. Gutierrez-Piñeres y G.A. González, International Journal of Theoretical Physics, **51**, 1737–1753 (2012)

Chapter 15
Tomographic Represention of Quantum and Classical Cosmology

Cosimo Stornaiolo

Abstract In this paper we consider a tomographic representation of quantum cosmology, in which tomograms (i.e. a standard positive probability distribution function) describe the quantum state of universe in place of the the wave function or density matrices. This representation can be extended to classical cosmology and used to reconstruct the initial conditions of the universe by studying the evolution of a quantum tomogram to a classical one. To this end we give a definition of classical tomogram based on cosmological observations, and we give the criterion for the reconstruction of the states primordial universe.

15.1 Introduction

Quantum cosmology [1] is an application of quantum gravity in which the full field theory is reduced to a problem with a few degrees of freedom, by the restriction of the superspace, i.e. space of the spatial metrics, to the so-called minisuperspace, which is the space of the homogeneous metrics. In this case the quantum properties of homogeneous cosmological models can be described in terms of quantum mechanics. Quantum cosmology can be considered as a toy model designed to capture some of the fundamental properties of the complete field theory, even, if by fixing contemporarily most of the field modes and of their conjugate momenta to zero, it violates the uncertainty principle. There is a departure from the Copenhagen interpretation where the measurements are conceived by taking a quantum system embedded in a classical space, where some classical observer makes the measurements. By considering the whole universe as a quantum system, we necessarily have a non conventional approach to a quantum theory, because we are dealing with the entire universe as if it were a single particle. Unlike a microscopic system, in which generally are considered ensembles of many particles, the theory is not able to make predictions on the statistical evolution of the system. However, a phenomenological approach to the theory and how quantum effects affected the background radiation and on the galaxy distribution was recently proposed (see e.g. [2, 3]). Also our work

C. Stornaiolo (✉)
Istituto Nazionale di Fisica Nucleare, Complesso Universitario di Monte S. Angelo, Sezione di Napoli, Napoli 45–80126, Italy
e-mail: cosmo@na.infn.it

C. Moreno González et al. (eds.), *Accelerated Cosmic Expansion*, Astrophysics and Space Science Proceedings 38,
DOI 10.1007/978-3-319-02063-1_15,

proposes a phenomenological approach to quantum cosmology, on different lines, but not in contrast with those mentioned before.

In quantum cosmology one first introduces a canonical formalism and then defines a wavefunction of the universe which evolves according to the Wheeler-DeWitt equation. It is necessary to specify the initial conditions for the wave function in order to determine the evolution of the universe. They must be considered as a fundamental law of physics [4], because differently from any other physical system, where the initial conditions can change from one configuration to another, they determine the evolution of the whole universe, whose configuration is once and for all. Theoretical approaches to this law have been posed by Hartle-Hawking [5], Vilenkin [6] and Linde [7]. But as for many physical laws, we believe that a phenomenological approach to reconstruct the initial conditions of the universe is possible. A limiting condition to the reconstruction of the initial conditions could be the huge amount of entropy of the universe $S/k \sim 10^{80}$, which measures the lost of informations. But this number is significantly small when compared with the maximum entropy which, according to the holographic principle, is of the order of $S/k \approx 10^{120}$. This guarantees that the initial conditions of the universe can be reconstructed with a very high accuracy. But in order to do this we need to connect the cosmological observations available in the present time to the early state of the universe. This will be discussed in the following of the paper. Finally it is important to have a mechanism to describe the transition from quantum physics to classical physics by means of the quantum decoherence. Not all the solutions of the Wheeler-DeWitt lead to the decoherence, the so-called Hartle conditions are the necessary conditions needed to guarantee this transition.

In this paper we present an alternative representation of quantum cosmology by introducing the notion of standard positive probability distribution function or tomogram which has been used [8] to describe the quantum state of universe alternatively to the wave function or to the density matrix descriptions. Tomograms where introduced in quantum mechanics in analogy to the same notion used in quantum optics in [9]. The advantage of using the tomographic representation of quantum mechanics is that one can represent a quantum state by an observable function. Moreover even if we quantum tomograms instead of the wavefunction, it is also possible to define classical tomograms. Classical and quantum tomograms are both defined on the phase space and it is possible to show that quantum tomograms can evolve to classical ones. So one can think to reconstruct the initial conditions of the universe by first introducing a classical tomogram of the universe from observations and then one can reconstruct the initial state of the universe projecting back in time the this tomogram.

In this following sections we introduce first the tomographic representation of quantum mechanics and then apply it to quantum cosmology. We obtain the WheelerDeWitt equation for tomograms (TWDW equation), the probability transition function which defines the evolution of a tomogram and finally we give a prescription for deriving a classical tomogram for the present time based on a statistical function distribution of the distance and velocity of the galaxies, in doing so we slightly modify the paradigm of quantum cosmology, which traditionally assumes a rigorous homogeneity of the universe. Since the TWDW equation is not time dependent, we discuss the meaning of the time evolution of the tomogram as the evolution of the observations by a classical observer in the late stages of the cosmological evolution and by a quantum observer in the early universe.

15.2 Tomographic Representation of Quantum Mechanics

To understand better the role of the tomograms in reconstructing the state of a quantum system, we can refer to an experiment done by Kurtsiefer, Pfau and Mlynek [10], where a coherent beam of helium atoms in a double-slit experiment measurements of the quantum mechanical analogue of the classical phase space distribution function show that the motion of atoms behaves in a strongly non classical manner. The experiment was designed to reconstruct the Wigner function of the ensemble of the helium atoms by means of time-resolved diffraction patterns. These patterns are described by distribution functions $P(\tilde{x}, t_d)$ where t_d is the traveling time between the double-slit and the atom detector. It can be shown that this function is related to the marginal distribution $P_\Theta = P(\frac{\tilde{x}}{\cos\Theta}, \frac{mx_0^2}{\hbar}\tan\Theta)$, which is the corresponding tomogram. Finally the authors were able to reconstruct to the Wigner function, see Eq. (15.4). This paper therefore shows that tomograms are observable quantities which represent a quantum state. Therefore, according to [9] tomograms can be used in quantum mechanics alternatively to the wavefunctions.

Similarly, we introduce a tomogram in quantum cosmology instead of the wavefunction of the universe. We can also introduce the tomograms in classical cosmology and study directly the evolution of a quantum cosmological tomogram into a classical one or vice versa we can first construct from observations a classical tomogram and reconstruct the quantum tomogram of the early universe by using the transition probability functions, which are defined later.

To give the notion of tomogram let us first recall the definition of the Wigner function. It was introduced by Wigner to study quantum corrections to classical statistical mechanics and it can be assigned to represent a quantum state. It is expressed in terms of density matrix ($\hbar = 1$) by

$$W(q,p) = \int \rho\left(q + \frac{u}{2}, q - \frac{u}{2}\right) e^{-ipu} du \tag{15.1}$$

The inverse transform reads

$$\rho(x,x') = \frac{1}{2\pi}\int W\left(\frac{x+x'}{2}, p\right) e^{ip(x-x')} dp. \tag{15.2}$$

The Radon transform of the Wigner function in the modified form is the integral transform of the form

$$\mathcal{W}(X,\mu,\nu) = \int W(q,p) e^{ik(X-\mu q-\nu p)} \frac{dk\,dq\,dp}{(2\pi)^2} \tag{15.3}$$

Here X, μ and ν are real numbers. The Wigner function can be found using the inverse Radon relation

$$W(q,p)=\frac{1}{2\pi}\int e^{i(X-\mu q-\nu p)}\mathcal{W}(X,\mu,\nu)dXd\mu d\nu. \tag{15.4}$$

The standard Radon transform is obtained from the two above by putting $\mu=\cos\Theta$, $\nu=\sin\Theta$.

One can see that the tomographic symbol of density matrix is given as a marginal distribution since

$$\mathcal{W}(X,\mu,\nu)=\int W(q,p)\delta(X-\mu q-\nu p)\frac{dqdp}{2\pi} \tag{15.5}$$

It is clear that

$$\int\mathcal{W}(X,\mu,\nu)dX=1, \tag{15.6}$$

since the Wigner function is normalized

$$\int W(q,p)\frac{dqdp}{2\pi}=1 \tag{15.7}$$

for normalized wave functions.

The formulae (15.5) – (15.7) are valid for arbitrary density matrices, both for pure and mixed states. For pure states of the universe, the tomographic symbol can be expressed directly in terms of the wave function of the universe. The relation between the wave function and the tomogram is given by

$$\mathcal{W}(X,\mu,\nu)=\frac{1}{2\pi|\nu|}\left|\int\psi(y)e^{\frac{i\mu}{2\nu}y^2-\frac{iX}{\nu}y}dy\right|^2. \tag{15.8}$$

this relation is invertible. Tomograms and wavefunctions are in a one-to-one relation (except for a phase factor). This is why tomograms represent the quantum states of a system.

Notice that the formula relating the tomographic symbol with the wave function contains the integral

$$I=\left|\int\psi(y)e^{\frac{i\mu}{2\nu}y^2-\frac{iX}{\nu}y}dy\right| \tag{15.9}$$

In case of $\mu=0$, $\nu=1$ this integral is a conventional Fourier transform of the wave function. For generic μ, ν the integral is identical to the modulus of Fractional Fourier transform of the wave function [11]. Thus, the Radon transform of the Wigner function in the case of pure states is related to the Fractional Fourier transform of the wave function.

Classical tomograms are defined by replacing in Eq. (15.3) the Wigner function in Eq. (15.3) with any classical distribution function $f(q,p)$ in the phase space,

$$\mathcal{W}(X,\mu,\nu) = \int f(q,p)\delta(X-\mu q-\nu p)\frac{dqdp}{2\pi}. \tag{15.10}$$

The main difference between the classical and quantum tomograms is that the quantum tomograms must satisfy the inequality

$$\left[\int \mathcal{W}(X,1,0)X^2 - \left\{\int \mathcal{W}(X,1,0)X\right\}^2\right] \times \left[\int \mathcal{W}(X,0,1)X^2 - \left\{\int \mathcal{W}(X,0,1)X\right\}^2\right] \geq \frac{1}{4} \tag{15.11}$$

derived from the uncertainty principle. In conclusion formulae (15.5) and (15.10) show that quantum and classical tomograms are the set of all the probability distribution functions defined on the straight lines $X=\mu q+\nu p$ which span the whole phase space by varying μ and ν.

15.3 The Phase Space in Cosmology

Many homogeneous cosmological models can be derived from a point particle Lagrangian. For example for a Friedman-Lemaitre-Robertson-Walker universe with metric

$$ds^2 = -c^2dt^2 + \frac{a^2}{1-kr^2}\left(dr^2+r^2d\theta^2+r^2\sin^2\theta d\phi^2\right) \tag{15.12}$$

the Lagrangian is obtained by substituting the metric (15.12) into the general relativistic action $\int \sqrt{-g}R$

$$\mathcal{L} = 3a\dot{a}^2 - 3\,ka - 8\pi G\rho_0 a_0^{3\gamma} a^{3(1-\gamma)}. \tag{15.13}$$

The material part is given by the potential term $\Phi(a) = 8\pi G\rho_0 a_0^{3\gamma} a^{3(1-\gamma)}$, when the matter source is a fluid with equation of state $P=(\gamma-1)\rho$.

The tomogram has to be expressed in terms of the phase space coordinates which are cosmological observables. The expansion factor a is not an observable, instead it is the ratio $q=\frac{a}{a_0}=\frac{1}{z+1}$. Its conjugate momentum is $p_a=\frac{\partial L}{\partial \dot{a}}=6a\dot{a}$. The observable is $p=6\frac{a}{a_0^2}\dot{a}\equiv 6\frac{a^2}{a_0^2}H$

15.4 The Wheeler-DeWitt Equation and the Corresponding Equation for the Tomogram

The universe in a model of quantum cosmology is described by a wave functional which depends on the spatial metric. There exist several elaborated examples of minisuperspaces, here we consider, as an example, the model in which the metric dependence is reduced to dependence only on the expansion factor of the universe. This is a one dimensional Wheeler-DeWitt equation for a FLRW universe of the form

$$\frac{1}{2}\left\{\frac{1}{a^{\lambda}}\frac{d}{da}a^{\lambda}\frac{d}{da}-a^{2}+\Lambda a^{4}\right\}\psi(a)=0 \tag{15.14}$$

Here a, is the expansion term of the classical theory and λ is an index introduced to take into account the ambiguity of operator ordering. The Radon transform considered previously makes sense only for variables which take values from $-\infty$ to $+\infty$, then with the change of variables $a=\exp x$ and the Wheeler-DeWitt equation becomes

$$\frac{1}{2}\left\{\exp(-2x)\frac{d^{2}}{dx^{2}}+(\lambda-1)\exp(-2x)\frac{d}{dx}-2U(x)\right\}\Psi(x)=0 \tag{15.15}$$

where $U(x)=(\exp(2x)-\Lambda\exp(4x))/2$. This equation can be written also in the form [8]

$$\frac{1}{2}\left\{\exp(-2x')\frac{d^{2}}{dx'^{2}}+(\lambda-1)\exp(-2x')\frac{d}{dx'}-2U(x')\right\}\Psi^{*}(x')=0. \tag{15.16}$$

The TWDW equation corresponding to (15.15) is then

$$\begin{aligned}&\left\{\mathrm{Im}\left[\exp\left[2\left(\frac{\partial}{\partial X}\right)^{-1}\frac{\partial}{\partial\mu}+i\nu\frac{\partial}{\partial X}\right]\left(\frac{1}{2}\mu\frac{\partial}{\partial X}-i\left(\frac{\partial}{\partial X}\right)^{-1}\frac{\partial}{\partial\nu}\right)^{2}\right]\right.\\&+(\lambda-1)\mathrm{Im}\left[\exp\left(2\left(\frac{\partial}{\partial X}\right)^{-1}\frac{\partial}{\partial\mu}+i\nu\frac{\partial}{\partial X}\right)\left(\frac{1}{2}\mu\frac{\partial}{\partial X}-i\left(\frac{\partial}{\partial X}\right)^{-1}\frac{\partial}{\partial\nu}\right)\right]\\&-2\mathrm{Im}\left[\exp\left(-2\left(\frac{\partial}{\partial X}\right)^{-1}\frac{\partial}{\partial\mu}+i\nu\frac{\partial}{\partial X}\right)-\Lambda\exp\left(-4\left(\frac{\partial}{\partial X}\right)^{-1}\frac{\partial}{\partial\mu}\right.\right.\\&\left.\left.\left.+2i\nu\frac{\partial}{\partial X}\right)\right]\right\}\mathcal{W}(X,\mu,\nu)=0.\end{aligned} \tag{15.17}$$

15.5 Evolution of a Tomogram

The tomographic map can be used not only for the description of the universe state by probability distributions, but also to describe the evolution of the universe (quantum transitions) by means of the standard real positive transition probabilities. The transition probability

$$\Pi(X, \mu, \nu, t, X', \mu', \nu', t_0)$$

is the propagator which gives the tomogram of the universe $\mathcal{W}(X, \mu, \nu, t)$, if the tomogram at the initial time t_0 is known, in the form

$$\mathcal{W}(X, \mu, \nu, t) = \int \Pi(X, \mu, \nu, t, X', \mu', \nu', t_0)\mathcal{W}(X', \mu', \nu', t_0)dX'd\mu'd\nu'. \quad (15.18)$$

The positive transition probability describing the evolution of the universe has the obvious nonlinear properties used in classical probability theory, namely

$$\Pi(X_3, \mu_3, \nu_3, t_3, X_1, \mu_1, \nu_1, t_1) = \int \Pi(X_3, \mu_3, \nu_3, t_3, X_2, \mu_2, \nu_2, t_2)$$
$$\times \Pi(X_2, \mu_2, \nu_2, t_2, X_1, \mu_1, \nu_1, t_1)\, dX_2\, d\mu_2\, d\nu_2. \quad (15.19)$$

They follow from the associativity property of the evolution maps.

We stress the importance of Eq. (15.18), because it shows that given a tomogram at a time t_0, one can reconstruct the tomogram, i.e. the state of the universe, at any other time by applying this equation. The probability functions are derived according to the cosmological equations at each stage of the universe (i.e. inflation epoch, radiation and matter epochs). The total transition function $\Pi(X, \mu, \nu, X', \mu', \nu', t)$ is obtained by repeatedly applying Eq. (15.19) for the transition functions of each cosmological epoch.

Therefore, in order to reconstruct the initial state of the universe, we need to find a phenomenological tomogram to insert in Eq. (15.18), defined on the phase space with coordinates (assuming $a_0 = 1$)

$$p = 6a\dot{a} = 6a^2 H \quad \text{and} \quad q = a. \quad (15.20)$$

15.6 The Classical Tomogram

The Wheeler-DeWitt equation is a Schrödinger-like equation in which time is not present. In the example of Sect. 15.4, the wave function depended only on a or on $x = \ln a$. So is for its tomographic version the TWDW equation. We assume to take the solutions of these equations on the light-cone, because all the information of the state of the universe come only from the cosmological observations along the light-cone. Let us suppose for the moment that we can observe the universe at all values of z.

The next point is how to construct a tomogram as a statistical function. If the universe were perfectly homogeneous and isotropic, since we are considering a constrained system, the tomogram should be reduced to a point for each value of a. Since on the other hand the universe is not totally homogeneous, generally many cosmological observables, like the Hubble constant H_0 are averages of measures taken along all directions. To construct a realistic tomogram we adopt the point of view of [14] where the variance of the Hubble flow is considered without making a priori cosmological assumptions and is viewed as the differential expansion of regions of different local densities, because during the evolution of the universe they may have had different expansion histories from an epoch when the density was close to uniform. Therefore taking measuring the observables along sectors of the sky defined by different directions and a particular angle around each direction, we calculate their averages on these sectors. These averages form a statistical sample to be inserted on the phase space defined above. The classical tomogram will be the distribution function of points on each straight line passing through the origin. Each straight line will correspond some definite redshift z. The classical tomogram so-defined is given by the state of the universe at all the epochs, in line of principle it should be extended to the farthest parts of the universe in causal connection with us. But at present the current observations show distances up to $z = 8, 9$ using the standard candles, and at about $z = 1000$ from observations of the CMB radiation, which is a limitation to the actual construction of the tomogram.

What is the meaning of the time dependence of this tomogram? As tomograms are related to the observations, so they *define* also the observer. By observer we mean any physical system which records the informations coming from the universe. A tomogram is classical when there is a very weak interaction between the observer and the rest of the universe and new version of the cosmological principle is that all the observers see the same tomogram. On the other side the resulting tomogram is quantum when the "observer" interacts with the entire universe and it should be subject to the uncertainty conditions (15.11).

15.7 Conclusions

In this paper we showed that the reconstruction of the initial conditions of the universe can be done first by constructing an observable functions which can be related to the early stages of the universe by means of probability transition functions. We defined such a function by weakening the assumption of homogeneity of the universe and considering the variance of Hubble flow in different regions of the universe. A classical tomogram was then defined by considering observations at any red-shift until the maximal distance where events can be observed. Even if the definition, due to the limitations of current observations, is not viable from a practical point of view, it is in principle interesting, because it relates the state of the universe with the observer. A better definition would be to define the tomogram on a spatial hypersurface at a time t_0 and consider the transitions from one hypersurface to another. This is the aim of a future work.

References

1. D.L. Wiltshire, An Introduction to quantum cosmology, In *Canberra, Cosmology* 473–531 (1995) [gr-qc/0101003]
2. M. Bojowald, Quantum Cosmology: Effective Theory, Class. Quant. Grav. **29**, 213001 (2012) [arXiv:1209.3403 [gr-qc]]
3. C. Kiefer, M. Kraemer, Int. J. Mod. Phys. D **21**, 1241001 (2012) [arXiv:1205.5161 [gr-qc]]
4. J.B. Hartle, Quantum cosmology: Problems for the 21st century, In *Nishinomiya 1996, Physics in the 21st century* 179–199 [gr-qc/9701022]
5. J.B. Hartle, S. W, Phys. Rev. D **28**, 2960 (1983)
6. A. Vilenkin, Phys. Rev. D **33**, 3560 (1986)
7. A.D. Linde, Sov. Phys. JETP **60**, 211 (1984) [Zh. Eksp. Teor. Fiz. **87** (1984) 369]
8. V.I. Manko, G. Marmo, C. Stornaiolo, Gen. Rel. Grav. *37* 99 (2005); V. I. Man'ko, G. Marmo, C. Stornaiolo, Gen. Rel. Grav. **37** (2005, 2003); V.I. Man'ko, G. Marmo, C. Stornaiolo, Gen. Rel. Grav. **40** (2008, 1449); S. Capozziello, V. I. Man'ko, G. Marmo, C. Stornaiolo, Gen. Rel. Grav. **40** (2008, 2627); S. Capozziello, V. I. Man'ko, G. Marmo, C. Stornaiolo, Phys. Scripta **80**, 045901 (2009)
9. S. Mancini, V.I. Manko, P. Tombesi, Quant. Semiclass. Opt **7**, 615 (1995); S. Mancini, V.I. Manko, P. Tombesi, Phys. Lett **A213**, 1 (1996)
10. Ch. Kurtsiefer, T. Pfau, J. Mlynek, Nature **386**, 150–153 (1997)
11. V.I. Manko, R.V. Mendes, Phys. Lett. **A263**, 53 (1999)
12. O.V. Manko, V.I. Manko, G. Marmo, Phys . Scr. **62**, 446 (2000); J. Phys. A **35**, 699 (2001)
13. S.W. Hawking, in *Intersection between Elementary Particle Physics and Cosmology* eds. by T. Piran, S. Weinberg (World Scientific Publishing Co, Singapore 1986)
14. D.L. Wiltshire, P.R. Smale, T. Mattsson and R. Watkins, arXiv:1201.5371 [astro-ph.CO]

Index

C. Moreno González et al. (eds.), *Accelerated Cosmic Expansion*,
Astrophysics and Space Science Proceedings 38,
DOI 10.1007/978-3-319-02063-1, © Springer International Publishing Switzerland 2014

The manufacturer's authorised representative in the EU is Springer Nature Customer Service Centre GmbH, Europaplatz 3, 69115 Heidelberg, Germany. If you have any concerns regarding our products, please contact ProductSafety@springernature.com

Printed and bound by CPI Group (UK) Ltd, Croydon, CR0 4YY
30/09/2026
02225569-0004